Mediaeval Philosophical Texts in Translation

No. 32

Roland J. Teske, S.J., Editor

Henry of Ghent

QUODLIBETAL QUESTIONS

ON FREE WILL

Translated from the Latin

with an Introduction and Notes

by

Roland J. Teske, S.J.

Marquette University Press
Milwaukee, Wisconsin

Printed in the United States of America
ISBN 0-87462-235-2
Library of Congress Catalogue Card Number: 92-83749

To my colleagues

in the Department of Philosophy

at Marquette University

Acknowledgements

The existence of this volume in Mediaeval Philosophical Texts in Translation is largely due to the help and encouragement of the Reverend Raymond Macken, O.F.M., the coordinator of the critical edition of the works of Henry of Ghent, at the De Wulf-Mansion Centre at the Catholic University of Louvain (K. U. Leuven), Leuven, Belgium. I gratefully acknowledge his suggestions, inspiration, and encouragement. I began work on the translation of Henry's *Quodlibets* on human freedom with at most a mild interest and gradually came to see that Henry of Ghent had interesting and important things to say on human freedom. I hope that this volume will lead others to find in Henry a source of philosophical inspiration and reflection.

This volume also owes its existence to the generous help of my colleague and friend, Dr. Lee Rice, of the Department of Philosophy at Marquette, who had been assistant editor of this series for the last three volumes and who has contributed greatly to the present volume as well. To him I also express my deep gratitude and appreciation. Finally, I want to thank Mr. Steven Barbone, a graduate student in the Department of Philosophy, who has spent many hours of careful work in proofreading the text and in producing the camera-ready copy for the printer. His work on this and the other recent volumes of the series is deeply appreciated.

Table of Contents

Preface

As Coordinator of the *Henrici de Gandavo Opera Omnia* since its beginning, it is for me an honour and a joy to present the excellent work of Professor Roland J. Teske, S.J., Professor of Philosophy at Marquette University, Milwaukee, Wisconsin, U.S.A., which is entitled *Henry of Ghent: Quodlibetal Questions on Free Will*.

The publication of the volumes of the critical edition of Henry's *Opera Omnia* by the Catholic University of Leuven (Louvain) makes continual progress. Over the past thirteen years, thirteen volumes of the approximately forty-five anticipated volumes have already appeared. The great lines of this lofty medieval cathedral of philosophical-theological thought, critically edited in our modern time, become ever clearer, and a series of substantial parts of it have already been restored to the splendor of their original text.

Professor Teske is the first to use the critical edition of the *Opera Omnia* for offering the modern reader, in a great international language of our time, a volume of characteristic extracts from Henry's thought. Because Henry's thought, spread as it is over many volumes to be edited, remains on many points a *"mare incognitum,"* an uncharted sea, Professor Teske had the wisdom to limit himself to one important area of it, namely, the human will and its freedom. We hope that other volumes in great international languages will follow his excellent example for other areas of Henry of Ghent's world of thought.

We are convinced that the numerous intellectuals of our time who read the language of Shakespeare will immediately notice how, thanks to this book, Henry's daring, profound and lofty thought on human freedom is made accessible in a way valuable to the modern reader. Professor Teske's introduction gives in a clear and succinct manner the essentials of Henry's life and writing, sketches his Quodlibetal Questions, and then indicates the major lines of his teaching in them concerning the treated theme. He then directly confronts the modern searcher after truth with some carefully selected questions from these Quodlibets, translated in a faithful and at the same time clear and handy way. Henry presented these Quodlibets and all his other works in "the beautiful style of Platonism," amid his unremitting, continuous activity during his long and successful teaching careeer at the University of Paris in the last quarter of the thirteenth century. This book clearly sets forth the fundamental views of this great thinker of the past on this important area of philosophy and thus makes them available for a balanced creative philosophical synthesis during our time.

Indeed, by his wise initiative, Professor Teske has put Henry's deep and profound thought in this area into the hands of the creative thinkers and searchers after the truth of our time. Henry was always considered a great and independent thinker, especially loved and esteemed by those who, as he did, follow the more Platonic paths of the "Father of the West," Augustine, and are independent spirits themselves. We are convinced that the modern public will understand him. Thanks to the lucid explanations and the careful translations of Professor Teske, they will see that Henry was a prudent and balanced thinker who had as his great ideal to bring light to the spirit of other humans and warmth to their heart. Henry was not, like Augustine, a bishop, but he had the ideal of being, in his role as a medieval *magister* – as he expressed it in his first Quodlibet – "a great light in the Church."

But Henry would not do that work alone. The intellectual reader of our time will immediately remark that this quite sympathetic, rather Platonic thinker has very deep roots in the scholastic ideal of the *"philosophia perennis."* He consciously belongs to the solid and continuous philosophical tradition running back to the great Greek and Latin philosophers, such as Plato, Aristotle, Augustine, Seneca, Boethius, and Anselm, along with an Arab component, such as Avicenna. Dreams are often more practical than the daily actuality. This Western scholastic dream of a "philosophy which defies the centuries," deeply rooted in ancient thought, adjusted and enlarged in a systematic way by the medieval *Summae*, university disputations, and careful daily university teaching, continued and modernized in the Renaissance and modern times up to today, has become one of the greatest forces which make the actual Western civilization so helpful, efficient, and successful. Indeed, we possess, thanks to this secular work, a trustworthy instrument for the right organization of the life of the human person and of human society on the global scale.

"Habent sua fata libelli": Each book leads its own life. Our best wishes accompany this valuable work of Professor Teske. May it reach in the entire world, thanks to the vehicle of a great modern international language, its true public: the creative spirits and the searchers after the truth.

Dr. Raymond Macken
Coordinator
HENRICI DE GANDAVO *Opera Omnia*
Louvain, Easter 1992

Introduction

I. The Life and Writings of Henry of Ghent

Little is known with certainty about the life of Henry of Ghent. His name indicates that he was probably born in Ghent, though the date of his birth is unknown.[1] Toward the end of the nineteenth century German and French scholars stripped away a great amount of the legend that had grown up around Henry, for instance, that he was a member of the famous Goethals family of Ghent, a student of Albert the Great, a master first in Cologne, then in Ghent and finally in Paris, and a member of the Order of Servites.[2] It is true that in the beginning of the sixteenth century the Order of Servites adopted Henry as their official doctor on the belief that he had been a Servite and that this led to a new edition of his two principal works and some significant studies of the writings of the Solemn Doctor.[3]

Henry probably studied the liberal arts at the University of Paris, where he became master of theology in 1275. Since for that rank he should have been thirty five years old, his date of birth can be placed at 1240 or earlier. From a rare autobiographical reference in *Quodlibet* XIII, question 14, the general editor of Henry's *Opera Omnia*, Fr. Raymond Macken, O.F.M., concludes that Henry was present in Paris either in the summer or fall of 1264.[4] Furthermore, his first quodlibetal disputation was held at Christmas in 1276, with the second and third being held at Christmas in 1277 and at Easter in 1278. In the manuscripts of the latter two he is referred to as archdeacon of Brüges, and in the manuscript of

---

1. J. Paulus places the date of Henry's birth quite early, c. 1217, in his article on Henry in *The New Catholic Encyclopedia* (New York: McGraw-Hill, 1967), vol. 6, pp. 1135-1137, where he also states that Henry was a student of William of Auvergne, who died in 1249. On the other hand, in his book, *Henri de Gand. Essai sur les tendances de sa métaphysique* (Paris: J. Vrin, 1938), p. xiii, he simply states that it is impossible to determine the date of his birth.

2. For the most recent English-language study of Henry's life and works, see the "Introduction" to Steven Marrone's *Truth and Scientific Knowledge in the Thought of Henry of Ghent* (Cambridge, MA: The Medieval Academy of America, 1985), pp. 1-11. Cf. J. Paulus, *Henri de Gand*, p. xiii, for detailed references to what is known of Henry's biography, and pp. xxi-xxii for some of the sources of the legend that developed about Henry.

3. Cf. *Henrici de Gandavo Quodlibet I*, ed. R. Macken (Leuven: University Press; Leiden: E. J. Brill, 1979), pp. xxvii-xxviii, for comments on the early editions of the *Quodlibeta*.

4. Cf. *Henrici de Gandavo Quodlibet I*, ed. R. Macken, pp. viii-ix.

the fourth quodlibetal disputation he is called archdeacon of Tournai.[5]
While Henry certainly traveled between Paris and Tournai, he was
present in Paris for his fifteen quodlibetal disputations which date from
1276 to 1291 or perhaps 1292. His *Summa*, derived from his *Ordinary
Questions* (*Quaestiones ordinariae*), along with the *Quodlibetal Questions*
(*Quodlibeta*), were edited by Henry himself during his long teaching
career from 1276 to 1293.[6] Henry was a member of the theological commission set up by Etienne Tempier, the bishop of Paris, to investigate the
errors being taught in the Faculty of Arts which led to the condemnaton
of 219 propositions on March 7, 1277.[7] Paulus claims that Henry was an
active supporter of this condemnation of Latin Averroism and that he was
later so violently opposed to the mendicant orders that he was severely
reprimanded by the future pope, Boniface VIII, in 1290.[8] Henry died on
June 29, 1293.

Besides the *Quodlibeta*, which will be discussed below, Henry's
extant writings include: his *Summa*, or *Quaestiones ordinariae*, two sermons, a treatise on the dispute between the bishops and the Friars
(*Tractatus super facto praelatorum et fratrum*), an explanation of the first
chapters of Genesis (*Expositio super prima capitula Genesis*), a commentary on the *Physics* of Aristotle (*Quaestiones in Physicam Aristotelis*), and
a logical treatise, *Syncategoremata*. The latter two works are probably the
work of Henry, while the commentaries on Aristotle's *Metaphysics* and on
the *Book of Causes*, along with a treatise on penance and a book on
virginity, are less probably Henry's work.[9]

Henry referred to his *Summa* as *Quaestiones ordinariae*. The
"ordinary questions" resulted from the normal teaching activity of a master in theology, while the quodlibetal questions reflected the formal
disputations that were held twice a year and treated questions submitted
by the audience. Hence, the quodlibetal questions reflect the current
topics of interest in philosophy and theology, while the *Summa* represents
the more systematic development of a master's thought as he presented it

---

5. Cf. *Henrici de Gandavo Quodlibet I*, ed. Macken, p. ix.
6. Cf. *Henrici de Gandavo Quodlibet I*, ed. Macken, p. x. Macken has found corrections on
   manuscripts of the *Summa* and the *Quodlibeta* that stem from Henry's own hand. Cf.
   R. Macken, "Les corrections d'Henri de Gand à ses Quodlibet," *Recherches de
   théologie ancienne et médiévale* 40 (1973), 5-51; id. "Les corrections d'Henri de Gand à
   sa Somme," *Recherches de théologie ancienne et médiévale* 43 (1976), 55-100.
7. Cf. *Henrici de Gandavo Quodlibet I*, ed. Macken, p. x. For a partial, but readily accessible
   English translation of the condemned propositions, cf. *Philosophy in the Middle Ages*,
   ed. A. Hyman and J. Walsh, 2nd ed. (Indianapolis: Hackett, 1983), pp. 584-591.
8. Cf. "Henry of Ghent," in *The New Catholic Encyclopedia*, vol. 6, p. 1035.
9. Cf. R. Macken, "Henri de Gand," in *Dictionnaire d'histoire et de géographie
   ecclésiastiques* 136, c. 1134.

to his students.[10] Though Henry had intended to write a full *Summa*, he completed only the treatise on God.[11]

Henry's other works are less important. One of his two extant sermons, "Sermon in the Synod on the Monday after 'The Mercy of the Lord,' 1287" (*Sermo in synodo, feria 2a post Misericordia Domini, 1287*) has been edited and identified as the inaugural discourse for the national council of French bishops in 1287 on the Monday after the second Sunday after Easter.[12] Henry's "Sermon on the feast day of St. Catherine" (*Sermo in die festo Sanctae Catherinae*) was delivered before the masters and students of the University of Paris.

Macken points out that "The Treatise on the Dispute between the Bishops and the Friars" (*Tractatus super facto praelatorum et fratrum*) is a disputed question belonging to *Quodlibet* XII, though it was frequently omitted in the manuscripts due to its length.[13] Henry's *Explanation of the First Chapters of Genesis* (*Expositio super prima capitula Genesis*) contains a general introduction to Sacred Scripture, an introduction to Genesis and a commentary on the first three chapters of Genesis.[14]

Some characteristic doctrines of Henry's philosophy are the denial of a real distinction between essence and existence in creatures and the assertion of what he referred to as an intentional distinction. Henry, likewise, held that the spiritual powers of the human soul were not really, but only intentionally, distinct from the soul and from one another. He taught a highly spiritual view of the human person which was closer to Plato and Avicenna than to Aristotle and Aquinas.[15] He held a form of corporeity that tied the spiritual soul to the body, though in material beings other than humans he held the unicity of the substantial form. He also tried to emphasize the spirituality of the human intellect and will, stressing their active and autonomous character.[16]

---

10. Cf. *Henrici de Gandavo Quodlibet I*, ed. Macken, p. xxi.
11. Cf. *Summa* a. 21, where Henry indicates the intended structure of the rest of the work. The *Summa* was first edited in Paris in 1520; it has been photographically reproduced at St. Bonaventure, New York, in 1961.
12. Cf. K. Schleyer, *Anfänge des Gallicanismus im 13ten Jahrhundert. Der Widerstand des französischen Klerus gegen die Privilegierung des Bettelorden. Historische Studien* XIV. (Berlin, 1937).
13. Cf. *Henrici de Gandavo Quodlibet I*, ed. Macken, p. xxiii; the treatise has been edited as Volume XVII in the *Opera Omnia*.
14. Macken has done the critical edition of this work; it has been reprinted in Volume XXXVI of the *Opera Omnia*, 1980.
15. Cf. Armand Maurer, "Henry of Ghent and the Unity of Man," *Mediaeval Studies* 10 (1948), 1-20.
16. For general treatments of Henry's philosophy in English, cf. E. Gilson, *History of Christian Philosophy in the Middle Ages* (New York: Random House, 1955), pp. 447-454 and 758-763; J. Paulus, "Henry of Ghent," in *The New Catholic Encyclopedia*, vol. 6, pp. 1135-1137; E. Fairweather, "Henry of Ghent," in *The Encyclopedia of Philosophy* (New York: Macmillan, 1967), vol. 4, pp. 475-476. For more specialized studies, cf. the

II. The Quodlibetal Questions

Unlike the disputed questions (*quaestiones disputatae*), which formed a regular part of the university teaching, the quodlibetal questions were public disputations that took place twice each year, shortly before Christmas and shortly before Easter. It was a right and a duty of a master in theology to hold such quodlibetal disputations.[17] Anybody was permitted to submit a question on any philosophical or theological topic, and from this practice these disputations have taken their name.[18] If the disputation was recorded by a student, it was called a *reportatio*. Henry himself carefully edited his own quodlibetal questions. M. Grabmann has said of Henry's quodlibetal questions:

> The *quodlibeta* of the Solemn Doctor, truly the most valu-
> able quodlibetal work of Scholasticism, are a highly
> important, but as yet insufficiently studied source for a
> deeper historical understanding of the inner doctrinal
> opposition between Augustinianism and Thomistic
> Aristotelianism of the thirteenth century.[19]

Despite their alleged value, the *Quodlibeta* have only recently begun to receive the scholarly attention they deserve, principally as the result of the critical edition of Henry's works that has been undertaken by R. Macken, at the De Wulf-Mansion Centre of the University of Louvain. A number of studies of Henry's thought have begun to appear in English.[20] This volume offers an English translation of a selection of Henry's *Quodlibetal Questions* that present the heart of his teaching on human freedom.

III. Henry's Teaching on Human Freedom

Henry's first quodlibetal disputation was held at Christmas time of 1276, only a few months before Etienne Tempier, the bishop of Paris, condemned as erroneous 219 theses allegedly taught in the Faculty of Arts. This condemnation on March 7, 1277, just three years to the day

bibliography of this volume.
17. R. Macken, "Heinrich von Gent im Gespräch mit seinen Zeitgenossen über die menschliche Freiheit," *Franziskanische Studien* 59 (1977), 125-182, here 127.
18. Cf. E. Gilson, *History of Christian Philosophy in the Middle Ages* (New York: Random House, 1955), pp. 247-248.
19. M. Grabmann, "Berhard von Auvergne, O.P. (d. nach 1304), ein Interpret und Verteidiger des hl. Thomas von Aquin aus alter Zeit," in *Mittelalteriches Geistesleben* (Munich: M. Hüber, 1936), p. 557 (my translation).
20. The bibliography at the end of this volume attempts to list the more important studies of Henry's philosophy, especially those available in English.

after the death of Thomas Aquinas, had as one of its aims the defense of human and divine freedom against some more radical forms of Aristotelianism. Propositions 150 to 169 explicitly touch upon the freedom of the will.[21]

As bishop of Paris, Etienne Tempier solemnly excommunicated those who taught any of the condemned propositions or presumed to defend them in any way or even listened to those who defended them without reporting their defenders to the bishop or his chancellor within a week.[22] Wippel argues that the condemnation ultimately stemmed from the introduction into the Christian West of the wealth of non-Christian learning during the preceding century. While some, such as Albert and Aquinas, aimed at an harmonious synthesis of Greek and Arabian philosophy with the Christian faith, others accepted the non-Christian philosophy almost in its entirety, seemingly to the detriment of the faith. The latter "Radical Aristotelians" or "Latin Averroists," such as Siger of Brabant and Boethius of Dacia on the Faculty of Arts, provoked in turn a conservative reaction that reasserted the supremacy of theology over philosophy. This Neo-Augustinian movement included John Pecham, William de la Mare, and Henry of Ghent.[23] The condemnation ended the university career of Siger of Brabant and banished Giles of Rome from the theology faculty in Paris from 1277 to 1285. Among the condemned propositions there were some that were far from being heretical or even dangerous. In fact, some condemned theses had been taught by Thomas Aquinas. It was not until 1325 that prohibition of the condemned propositions was revoked insofar as it touched the teachings of Aquinas.[24]

Henry of Ghent was a member of the theological commission that drew up the 219 propositions which Tempier condemned and a Neo-Augustinian who was deeply concerned to defend Christian orthodoxy. Hence, he is deeply opposed to positions that would detract from human freedom by treating the will as a passive power that needs to be actualized by something else, whether by the heavenly bodies or by the object known by the intellect.

---

21. Cf. Roland Hissette, *Enquête sur les 219 articles condamnés à Paris le 7 mars 1277* (Louvain: Publications Universitaire; Paris: Vander-Oyez, 1977), pp. 230-263, for an examination of the meaning and origin of the condemned propositions.
22. Cf. John F. Wippel, "The Condemnations of 1270 and 1277 at Paris," *The Journal of Medieval and Renaissance Studies* 7 (1977), 169-201.
23. Cf. Wippel, "The Condemnations," pp. 195-196.
24. Cf. Wippel, "The Condemnations," pp. 200-201.

1. Quodlibet I: The Opening Salvo

In Henry's first *Quodlibet* there are six questions on human freedom; they reveal to us Henry as a philosopher taking a decisive stand on one of the important questions of his day. In *Quodlibet* I, question 14, the first of the questions on human freedom included in this selection, Henry maintains that the will is a higher power than the intellect, and in question 15, which is also translated here, he argues that intellectual knowledge must precede the act of the will.

While the question about which power is the higher might seem unimportant in itself, it was a question that marked a point of division between those who emphasized the role of the intellect and those who emphasized the role of the will. On the "intellectualist" side was St. Thomas and many Aristotelians, while on the "voluntarist" side were the Neo-Augustinians, a group which included Dominicans trained prior to Aquinas, the Franciscans, and most of the secular masters.[25] There were clearly some on the "intellectualist" side of the spectrum who posed a threat to human freedom by their teaching.[26] Henry of Ghent has been viewed as a forerunner of John Duns Scotus and labeled an extreme voluntarist, though Macken has argued that he is more correctly viewed as being closer to the center of the spectrum, having given the primacy to the will without depreciating the human intellect.[27]

In question 14 Henry faces four objections that aim to show that the intellect is the higher power. First, Aristotle claimed that practical reason is the first mover in things to be done by the will. Second, Augustine held that intellectual contemplation is more excellent than action. Third, Scripture maintains that it was in his reason that man was formed anew in the image of God. Fourth, the judgment of reason directs the will and, thus, reason is superior to the will.

---

25. Following Macken, I use the term "Neo-Augustinian" in the sense given it by F. Van Steenberghen in *La philosophie au XIIIe siècle* (Louvain: Publications Universitaires, 1966), pp. 495-500; cf. Macken, "La volonté humaine, faculté plus élevée que l'intelligence selon Henri de Gand," *Recherches de théologie ancienne et médiévale* 42 (1975), 23, n. 59.

26. It is, on the other hand, much harder to identify the authors of specific condemned propositions, though Siger of Brabant and Boethius of Dacia are probably the main intended targets. Cf. R. Hissette, *Enquête sur les 219 articles*, pp. 495-500, for his attempts to trace the sources of the propositions.

27. Cf. R. Macken, "L'interpénétration de l'intelligence et de la volonté dans la philosophie d'Henri de Gand," in *L'homme et son univers au moyen âge. Actes du Septième Congrès Internationale de Philosophie Médiévale*, ed. C. Wenin (Louvain-La-Neuve: Editions de l'Institut Supérieur de Philosophie, 1986), pp. 808-814, here 808. Macken refers to the work of A. San Cristóbal-Sebastián, *Controversias acerca de la voluntad desde 1270 a 1300* (Madrid, 1958), pp. 155-157, for the view that Henry is an extreme voluntarist.

Henry claims that the will is the first mover of itself and of everything else in the whole kingdom of the soul.[28] Henry notes that, since the powers and substance of the soul are hidden from us, we have to investigate them from what is better known to us, namely, their habits, acts and objects. He states the general principle that, if the habit, act and object of one power are higher than the habit, act and object of another power, the first power is higher than the second power.

Henry argues that the habit of the will is higher than that of the intellect, since the habit of the will is charity, while that of the intellect is wisdom. So too, the act of the will is loving, while that of the intellect is knowing. Henry points out that the superiority of loving over knowing can be seen from a comparison of the two acts and from a consideration of how these two acts perfect their subjects. First, the act of the will is superior to the act of the intellect, because "the will is the universal and first mover in the whole kingdom of the soul."[29] Thus the will commands the intellect to consider, to reason, or to deliberate about whatever it wills whenever it wills, while the intellect in no sense commands the will. At the heart of Henry's view of freedom lies the image of the will as king or emperor over all the other powers of the soul, which may serve as counselors or ministers to the will, but do not have the power to command. Thus the will is the autonomous monarch of the soul. Secondly, in loving the will is perfected by the beloved as the beloved exists in itself, while in knowing the intellect is perfected by the known as the known has being in the knower. Thus in loving God we become like God as God is in himself, while in knowing God a likeness of God comes to be in us in accord with our manner of knowing. Henry admits that it is better to know things inferior to the soul than to love them, but claims that this fact merely indicates that the intellect is superior to the will "in a certain respect" (secundum quid). The first goodness and the first truth are the primary objects of the will and the intellect. Hence, we should judge one power to be superior to another without qualification (simpliciter) in terms of how the acts of those powers with respect to those primary objects perfect their subjects.

---

28. R. Macken has pointed out that Henry's comparison of the soul to a kingdom in which the will has the royal power of command can be found in other thinkers in the Neo-Augustinian tradition, such as Walter of Bruges. Cf. R. Macken, "Heinrich von Gent im Gespräch mit seinen Zeitgenossen über die menschliche Freiheit," *Franziskanische Studien* 59 (1977), 125-182, here 128, n. 16. See also my "The Will as King over the Powers of the Soul: Uses and Sources of an Image in the Thirteenth Century," *Vivarium*, forthcoming.

29. See *Quodlibet* I, q. 14, p. 26. This reference and subsequent references in this format refer to pages of the English translation that follows.

In his *Summa of Theology*, Thomas Aquinas had asked whether the will is a higher power than the intellect and concluded that intellect is the higher.[30] Thomas argues only from the objects of the two powers and points to the greater simplicity and abstractness of the object of the intellect as proof that the intellect is the higher power without qualification. He admits, nonetheless, that the will is at times higher than the intellect in a certain respect, for example, in relation for God, for love of God is better than knowledge of God. Macken points to the abstractness of Aquinas and claims that Henry aims at concreteness and a consideration of the two powers as directed to their principal object, namely, God.[31] A more basic difference lies in their divergent views of the human being and of human knowing. While Henry, following Avicenna, held a highly spiritual view of the human being, Thomas Aquinas emphasizes the substantial unity of the human person. While Henry speaks of the first truth as the primary object of the intellect, Thomas Aquinas stresses that the proper object of the human intellect in this life is not God, but the quiddity or nature existing in corporeal matter.[32] Finally, Henry argues that the will is superior to the intellect because the object of the will is the good without qualification, while the object of the intellect is truth, which is the good of the intellect and, thus, a good only in a certain respect. As such, truth is subordinated to the good without qualification, which is the ultimate end, the object of the will.

In dealing with the objections, Henry's responses are quite straightforward. His response to the second objection, however, is worth noting, for he claims that "to move" has two senses. In a metaphorical sense practical reason moves the one who wills – and not, properly speaking, the will – by proposing a goal toward which one should move. In another and truer sense, one thing moves another by acting upon it and compelling it to act; in this way the will moves the reason. Thus Henry attributes only a metaphorical causality to practical reason in proposing a goal and views motion in the true sense as a compelling efficient causality. In that sense, he cannot allow reason to "move" the will and still preserve the will's freedom. To be moved in the proper sense is to be passive and to be acted upon, while freedom can only be found in an active potency and in action.

In question 15 of *Quodlibet* I, Henry asks whether the act of the intellect precedes the act of the will. The single argument from an objector aims to show that the act of the intellect must precede that of the

---

30. Thomas Aquinas, *The Summa of Theology* (*Summa theologiae*) I, q. 83, a. 3.
31. R. Macken, "La doctrine de s. Thomas concernant la volonté et les critiques d'Henri de Gand," in *Tommaso d'Aquino nella storia del pensiero. Atti del Congresso Internazionale Roma-Napoli 17-24 aprile 1974* (Napoli, 1976), vol. II, pp. 86-87.
32. Thomas Aquinas, *The Summa of Theology* (*Summa theologiae*) I, q. 84, a. 7 c.

will, while the argument to the contrary tries to show that the desire to know, which is an act of the will, precedes the principles of knowledge and, hence, knowledge of everything else.

Henry answers that, if one is speaking about any knowledge whatsoever, the will cannot will anything without previous knowledge. Henry appeals to Augustine's argument that something of which we have absolutely no knowledge cannot be the object of the will. He then shifts the question to ask whether we can will something with only sensitive knowledge coming before the act of the will. Henry maintains that, in things which are moved by appetite and not merely by natural inclination, not only must some knowledge precede the appetition, but a particular sort of knowledge must precede a particular sort of appetition. Thus in human beings rational knowledge must precede the act of the rational appetite. Hence, intellectual knowledge must precede the action of the will.

Against the opposing argument, Henry claims that one does not proceed to a study of the principles of knowledge from a desire to know without having some general knowledge that stirs the will. Using the image of master and servant, he claims that the intellect as a servant directs the will as its master. Here the director is inferior, because the will can withdraw the intellect from its acts of directing and knowing as it wills.

2. Quodlibet IX: The Battle Resumed

Henry's ninth quodlibetal disputation was held at Easter in 1286.[33] It had been almost ten years since Henry last addressed the question of human freedom at any length. After the condemnation of 1277, the radical Aristotelians, such as Siger of Brabant and Boethius of Dacia, disappeared from the university scene, while Giles of Rome, a student of St. Thomas, was prevented from attaining the position of master until 1285 or 1286 and then only by the intervention of the pope and after the retraction of several of his theses. At Easter in 1286, the ninth question of the second *Quodlibet* of Godfrey of Fontaines marked a renewal of the intellectualist views with his extreme Aristotelian position on the supremacy of the intellect over the will.[34]

---

33. Cf. *Henrici de Gandavo Quodlibet I*, ed. Macken, p. xvii, who follows the dating of J. Gómez Caffarena, "Cronologia de la 'Suma' de Enrique de Gante por relación a sus 'Quodlibeta,'" *Gregorianum* 38 (1957), 116-133, here 116.
34. Cf. R. Macken, "Heinrich von Gent im Gespäch mit seinen Zeitgenossen über die menschliche Freiheit," *Franziskanische Studien* 57 (1977), 140-142. Cf. also O. Lottin, *Psychologie et morale aux xiie et xiiie siècles* (Gembloux: J. Duculot, 1957), vol. I, pp. 304-305.

In *Quodlibet* IX, question 5, Henry asks whether the will moves itself. His answer that the will does move itself runs directly against the basic Aristotelian principle that whatever is moved is moved by another.[35] Johann Auer claims that this point marks "the profound difference between the Aristotelian-Thomistic and the Scotistic-Franciscan conception of willing and freedom."[36] The objection raised against the will's moving itself is that the will would have to be in act in order to move and in potency in order to be moved, that is, in act and in potency at the same time. Since, moreover, the will is simple, it would have to be in act and in potency in the same respect.

Henry's answer is developed in a veritable treatise on motion in which he looks at motion at six levels, in which there is increasing difference between the mover and the moved as one moves down the scale of movers and things moved.[37] At the highest level, in the divine will, there is only a rational distinction between the mover, the moved, and the motion, that is, between the divine will considered as moving, the divine will considered as moved, and the divine act of willing. In God all three are really identical, though we speak of motion in the widest sense and extend its meaning to apply to what is utterly without motion.

In everything other than God there is a difference between the mover and the moved, the difference being greater to the extent that the mover and the moved are more distant from God. At the second level there is the motion in the human will, and at the third level that in the human intellect. At the fourth level there are heavy and light things that move by themselves to their natural place. At the fifth level there is the self-motion of animals, and at the sixth level there is the motion by which one thing brings another into being.

At the sixth level, at which new things are brought about or generated, the cause that moves is distinct from the effect produced in substance and essence, as well as in location and magnitude. For instance, the mother cat and her kitten are, at least at the end of the process of generation, distinct substances and distinct in place and mass. On the other

---

35. Cf. Aristotle, *Physics* VII, c. 1, 241b14-15 and VIII, c. 5, 257b2-13. This principle, which lies at the heart of the first way of Thomas Aquinas in *The Summa of Theology* (*Summa theologiae*) I, q. 1, a. 3, will accordingly prove useless to Henry for establishing the existence of God. Cf. Anton Pegis, "Henry of Ghent and the New Way to God (III)," *Mediaeval Studies* 33 (1971), 158-179.

36. J. Auer, *Die Entwicklung der Gnadenlehre in der Hochscholastik* (Freiburg: Herder, 1951), p. 145 (my translation). Cf. John Wippel, "Godfrey of Fontaines and the Act/Potency Axiom," *Journal of the History of Philosophy* 11 (1973), 299-317, for a defense of the Aristotelian principle by one of Henry's opponents.

37. Cf. R. Macken, "Der geschaffene Wille als selbstbewegendendes geistiges Vermögen in der Philosophie des Heinrich von Gent," in *Historia Philosophiae Medii Aevi. Studien zur Geschichte der Philosophie des Mittelalters. Festschrift für Kurt Flasch zum 60. Geburtstag*, ed. B. Mojsisch and O. Pluta (Amsterdam: Grüner, 1991), pp. 561-572, for an analysis of this question.

hand, at the fifth level of an animal's moving about locally, the mover and the moved are only partially different in substance and partially different in place. Henry's discussion of the local motion of animals, whether of those that walk or those that crawl or fly, relies heavily upon Aristotle's short treatises, *The Motion of Animals* and *The Forward Motion of Animals*. Though Henry is a Neo-Augustinian, he certainly takes pride in his knowledge of the Aristotelian text. So too, he makes great use of Averroes, known to the scholastics simply as the Commentator, because of his allegiance to Aristotle's thought. In noting that Averroes did not have a translation of these Aristotelian books, Henry implies that he is better versed in the Aristotelian text than Averroes.[38] After a lengthy discussion of animal self-motion, Henry concludes that in such local motion

> the mover and the moved are not distinct in their entirety and do not act upon each other from the outside. Rather, the soul is the same in the whole body, and it is utterly simple in the case of the intellective soul, although it is extended throughout the parts of the body in brute animals. Though it is indivisible, it organically moves the whole animal at a single point .... it moves itself in a sense insofar as it is in the parts it moves, and it also is moving insofar as it is in the point mentioned. But it moves itself accidentally ....[39]

That is, the soul of the animal that moves itself moves itself in much the same way that a helmsman moves himself in moving his ship. The long excursus on the motion of animals goes a long way toward explaining why his contemporaries referred to him as *Doctor digressivus* – the master of digression.[40] On the other hand, Henry's presentation of the Aristotelian account of how the animal's appetite is necessarily moved by the known sensible good clearly implies that the will cannot be so moved by the intellectually known good without the complete loss of freedom.

At the fourth level Henry places heavy and light things which move by themselves to their natural places. Here he considers inanimate bodies that are moved, not by appetition, but by natural impulse. The mover and the moved at this level are less distinct than at the previous level. Unlike animals, these inanimate bodies cannot begin moving or stop moving by themselves, but if they are outside their natural place and not impeded by something else, they move to their natural place as a result of

---

38. Cf. *Quodlibet* IX, q. 5, p. 34.
39. *Quodlibet* IX, q. 5, p. 45.
40. J. Paulus, "Henry of Ghent," in *The New Catholic Encyclopedia*, vol. 7, p. 1035. He was also known as *Doctor solemnis, Summus doctorum*, and *Doctor reverendus*.

their form. Heavy and light things, then, move themselves as wholes to their natural places. They are composed of matter and form, and form moves insofar as it is form, while it is moved insofar as it is in matter. Hence, the mover and the moved at this level are less different than at the level of animals where there are distinct organs.[41]

At the third level we have the mover and the moved in the intellect's act of understanding. The intellect is first moved by the intelligible object to the act of simple understanding. Then, by the power of conversion to itself, its act and its object, it opposes itself to itself as an intellective power naturally able to be moved by itself. The intellect moves insofar as it is informed with the knowledge of simple intelligence; the intellect is moved insofar as it is bare intellect and in potency to explanatory knowledge.[42] The intellect cannot bring itself into first act, but it can bring itself into second act once it has the knowledge of simple intelligence. Henry's treatment of the intellect is very brief and none too clear. Later, in contrasting the self-motion of the will with that of the understanding, he puts it more clearly.

> In the same way, the intellect is passive with respect to the object of simple understanding, but once it has been acted upon, it can act upon itself to generate in itself explanatory knowledge and to constitute new complex intelligibles regarding the intelligibles first known. So too, insofar as the intellect is in act with regard to the knowledge of principles, it brings itself into act with respect to knowledge of conclusions.[43]

Thus, once the intellect has been moved by the object to the knowledge of simple understanding, it can move itself to knowledge of complex intelligibles and can move itself from the knowledge of principles to that of conclusions.

When Henry comes to the mover and moved at the second level, he begins his discussion with a series of objections. First, some claim that the first motion of the will comes, not from the will itself, but from God. He mentions Anselm and Aristotle, but Thomas Aquinas, whom he does

---

41. For the discussion at this level Henry follows closely Averroes' commentaries on Aristotle's *Physics* and *The Heaven*, once again claiming for himself, it would seem, the right to be called an Aristotelian.
42. I have translated "*declarativa*" as "explanatory." Cf. Gómez Caffarena's *Ser Participado y Ser Subsistente*, p. 57, n. 51, where he cites Henry's *Summa* a. 40, q. 7, where Henry speaks of the definition as explanatory of what is known in simple intelligence and demonstrative knowledge as explanatory of complex knowledge.
43. *Quodlibet* IX, q. 5, p. 53.

not mention, had cited the relevant passages in *The Disputed Questions on Evil*.[44] Henry admits that God moves everything in his general governance of the world, but moves individual things according to their different characters. He seems to have held that the will can move itself from potency to act without being moved by God – a position that would certainly lessen or remove any danger to human freedom from the divine concurrence with the will of the creature.[45] Of this "virtual power" of the will, Macken says,

> To produce and specify their acts, spiritual faculties have only to develop their own virtuality without being determined to this by a cause that moves them from the outside. This virtuality (*virtus ad movendum*) corresponds to the substantial form of the subject of the spiritual faculty, and the subject has received this substantial form from an external productive cause (*generans*). In this sense the act which it produces can be called a 'virtual' act.[46]

Secondly, Henry mentions that some claim that the will is moved by the known good as a passive potency is moved by its proper cause. He rejects this view because it would mean that in the presence of the known good the will could not fail to be moved to willing. Thus free choice would be destroyed, as well as merit, deliberation, and the requisites of virtue.

Thirdly, Henry turns to the position of those who say that the form of the intellect is of itself the principle of human actions and that the will is merely an inclination following upon that form. Henry seems to allude to the view of St. Thomas who claimed that the known form is universal and not determined to only one action, just as the mind of a carpenter is not determinately inclined to building a particular house.[47] Henry argues that, if the form of the intellect moved the will by inclining it, then, just as the known universal good would move it indeterminately, the known particular good would move it determinately so that free choice would be destroyed. Henry mentions others who consider the act

---

44. Cf. *Quaestiones disputatae de malo*, q. 6 and q. 7, a. 1, arg. 8, where Thomas cites Aristotle, *Eudemian Ethics* VII, 14, 1248a24-27 and Anselm, *Why God Became Man* (*Cur Deus homo*) I, c. 11.

45. Henry "solves" the problem by, it seems, denying any need for God's antecedent concurrence with human willing. Cf. J. Owens, *An Interpretation of Existence* (Milwaukee: Bruce, 1968), pp. 113-120, for an account of St. Thomas's doctrine of God's concurrence with human freedom.

46. R. Macken, "Liberté humaine dans la philosophie d'Henri de Gand," in *Regnum Hominis et Regnum Dei. Acta Quarti Congressus Scotistici Internationalis. Studia Scholastico-Scotistica* 6 (1978), 577-584, here 579.

47. Cf. Thomas Aquinas, *The Disputed Questions on Evil* (*Quaestiones disputatae de malo*), q. 6.

of the will in terms of the determination of the act and in terms of the exercise of the act. Once again, he seems to be alluding to St. Thomas's distinction between freedom of specification and freedom of exercise.[48] According to this view, the intellect moves the will in terms of the specification of the act, but the will remains free in the exercise of its act, that is, free to will or not will the known good.[49] With regard to this distinction, Henry asks about the nature of the specification the will receives from the intellect. Either the intellect merely shows or offers the good to the will, or the intellect produces some inclination in the will. On the first alternative, the will is not moved by the known good or by the intellect; hence, if the will is moved, it is moved by itself both with regard to its specification and with regard to its exercise. According to Henry, the intellect, in showing or offering the object to the will, acts merely as an accidental cause and a necessary condition (*sine qua non*).[50] On the second alternative, some inclination is produced in the will. Either that inclination is some impression that inclines the will like a weight, as a habit might incline it, or it is the act of willing. Henry argues that, if the inclination is an impression like a weight inclining the will, the will still remains completely free to act in accord with or in opposition to that inclination. Hence, if it is moved to act, it is moved by itself both in its specification and in its exercise. If, on the other hand, the inclination is the act of willing, then, once again, the intellect cannot move the will to the specification of the act without moving it to the exercise of the act. Hence, Henry concludes,

> if the will were moved by the object of intellect however slightly, there could be no act of rejection concerning it. Rather, it would be necessary to carry out the act or to pursue the object to attain it.[51]

Henry indicates what he takes to be Aristotle's position regarding the will. According to Henry, Aristotle thought that the will is moved to the act of willing by the known object of desire, just as the intellect is moved to the act of assenting by the known truth. Indeed, just as the force of a demonstrative conclusion moves the intellect in speculative matters, so it moves the will in matters of action.[52] The Aristotelian contemporaries of Henry maintained that the will is, nonetheless, not moved by

---

48. Cf. Thomas Aquinas, *The Summa of Theology* (*Summa theologiae*) I-II, q. 9, a. 1 ad 3um, and *The Disputed Questions on Evil* (*Quaestiones disputatae de malo*), q. 6.
49. Henry does not find that this distinction of St. Thomas is of any help. Cf. Macken, "Heinrich von Gent," p. 147.
50. Henry of Ghent, *Quodlibet* IX, q. 5, p. 52.
51. *Quodlibet* IX, q. 5, p. 53.
52. *Quodlibet* IX, q. 5, p. 54.

violence or coercion, because it is not moved contrary to its nature. Henry insists, however, that "they completely remove freedom of choice in willing the object of the will, because it requires freedom from all necessity."[53]

Even if the good to be willed is determined by the intellect as the conclusion persuading one to will it or to deliberate about it, Henry's opponents would have to hold that the will moves itself to will to deliberate or is moved by something else to do so. Since they do not admit that the will moves itself, there must be another motion that moves the will to will to deliberate. Thus they are launched on an infinite regress, or they must come to a first mover of the will that is other than the will.[54] Aristotle attributed this motion ultimately to God, though more proximately to the sensible things which cause our knowledge and are determined by the heavenly bodies. Hence, Henry holds that it is false that the will is only moved by deliberation regarding means to an end. He states,

> without any deliberation determined by reason to one alternative, [the will] can move itself toward any good proposed, short of the last end when it is clearly seen. It can do so without movement from anything else, just as it can turn aside from it . . . .[55]

How then does the will move itself? Since the will is higher than the intellect and higher than all the other movers we have considered, there should be less of a difference in the will between the mover and the moved, though more than a merely rational difference as there is in God. Henry maintains that the potency in the will which receives the act of willing and the freedom of that potency do not differ merely by a rational distinction, but as powers of that potency. Henry finally comes to a direct answer to the question posed and states that

> we must reply to the question that the will alone moves itself to the act of willing . . . . It moves itself freely to the highest good seen as present, and through free choice to anything else. It also moves itself freely to the highest good, when it is seen as present only in the universal, just as in the universal no one cannot will to be happy.[56]

---

53. *Quodlibet* IX, q. 5, p. 55.
54. *Quodlibet* IX, q. 5, p. 56-57.
55. *Quodlibet* IX, q. 5, p. 59.
56. *Quodlibet* IX, q. 5, p. 64.

The will has such an active power over itself, because it can return to itself, like the intellect, insofar as it is separated from matter.

In turning to the objections, Henry admits that it is impossible for one and the same thing to be both in act and in potency at the same time and in the same respect. Hence, there must be some difference between the will as mover and the will as moved. Henry says,

> But in the present case the mover and the moved differ only by a distinction of reason and also by an intentional distinction. They do not differ as distinct potencies, but as powers of one potency.[57]

Thus Henry appeals to his "intentional" distinction to account for the difference between the will as mover and the will as moved.[58] Such a distinction is neither merely a distinction in the way we think about the will nor a distinction between two things. Henry's best known use of this distinction is found in his discussion of the distinction of being (*esse*) from essence (*essentia*) in creatures.[59]

The argument to the contrary should be conceded, Henry notes, because the conclusion is correct. On the other hand, he points out that the minor premise concerning the motion of animals and of light and heavy things is not relevant to the discussion of the created will.

In *Quodlibet* IX, question 6, Henry asks whether commanding (*imperare*) is an act of the will or an act of reason and intellect. The single objection claims that to command is merely to indicate to another that something should be done and that this is the function of reason.

Aquinas had held that to command is a function of reason, because it is the function of reason to direct toward an end.[60] In itself the question might seem relatively insignificant, especially since Henry did not hold a real distinction between reason and will, but the answer has significant repercussions elsewhere in moral philosophy and philosophy of law. For example, Aquinas defines law as a command of reason and would thereby exclude from moral theory anything like a theological voluntarism. So too, if a civil law were not the product of reason, but

---

57. *Quodlibet* IX, q. 5, p. 66.

58. Gómez Caffarena traces Henry's "intentional distinction" to Avicenna and likens it to Scotus's "distinction from the nature of the thing." Cf. *Ser Participado y Ser Subsistente en la Metafísica de Enrique de Gante* (Rome: Gregorian University Press, 1958), p. 91.

59. Cf. J. Paulus, *Henri de Gand. Essai sur les tendances de sa métaphysique* (Paris: J. Vrin, 1938), pp. 199-258, for a detailed discussion of the various distinctions in Henry, their sources and significance; cf. John Wippel, "Godfrey of Fontaines and Henry of Ghent's Theory of Intentional Distinction between Essence and Existence," in *Sapientiae procerum amore: Mélanges médiévistes offerts à Jean-Pierre Müller O.S.B.*, ed. Theodor W. Köhler (Rome, 1974), pp. 289-321.

60. Cf. Thomas Aquinas, *The Summa of Theology* (*Summa theologiae*) I-II, q. 17, a. 1.

merely the will or whim of the ruler, it would not, according to Aquinas, deserve to be called law.

Henry's argument to the contrary once again claims that the will has the power of command because it is supreme and free and possesses the greater dominion in the whole kingdom of the soul.

In answer to the question Henry grants that commanding is an act directed to someone in order to carry out something. He claims that, in order to determine to which power the act of commanding belongs, one must examine three factors: the relation of the one commanding him to whom the command is directed, the condition of the act commanded, and the disposition of the one to whom the command is given. Each of these considerations indicates that commanding is an act of the will, not of the intellect.

First of all, the one commanding must be superior to the one commanded, since an equal has no power to command an equal, nor does an inferior have the power to command a superior. But the highest power in a human being has command or dominion over the others. The only question is whether the intellect or the will is the highest power in a human being. Henry states the position of Thomas Aquinas that to command is the function of reason, but rejects it with an appeal to his previously argued position that the will is the higher power.

> For the will can will, even contrary to the dictate of reason, and can force reason to depart from its judgment and thus to agree with it, and it can constrain all the other potencies by its power of command.[61]

Admittedly, an act of the intellect must precede the act of the will, but the intellect determines for the will what it should will by something less than a command, because the will cannot be constrained by the intellect. Yet, such constraint is precisely what must be found in the one to whom a command is given. Henry allows that the intellect has a power of enjoining through persuasion and counsel, but lacks the power to compel that the will has. Following St. John Damascene, Henry maintains that the will can command the intellect and all other powers of the soul, but that the intellect cannot command the act of the will or any other act.

Second, one can see that the will has the power of command, from the condition of what is commanded, along with the relation of the one giving the command to the one to whom the command is given. As the one giving the command should be superior, so the commanded act should be an act of obedience. At times the intellect cannot obey the will,

---

61. *Quodlibet* IX, q. 6, p. 68.

if, for instance, the will commands something beyond its power, such as to understand a supernatural truth or to disagree with a clearly demonstrated truth. So too, the will cannot command the vegetative powers of the soul. Moreover, the sensitive appetites are at times not in full obedience to the will which has only political rule over them, not a despotic rule, as Aristotle pointed out in the *Politics*.[62] Henry holds that, since freedom of choice is not bound to obey any other power and since all other powers are bound to obey it, freedom of choice belongs to the will alone. He rejects St. Thomas's claim that "the root of freedom, as a cause, is reason or intellect,"[63] holding that the will is both the subject of freedom and its first root. Though the will derives its rationality from reason, it does not derive from reason its freedom of choice.

Following John Damascene, Henry allows one to ascribe freedom of choice to the actions of both reason and will, provided that one understands that such freedom is essentially in the will and only by participation in the intellect. So too, rationality can be attributed to the actions of both intellect and will, but it belongs essentially to the intellect and only by participation to the will.

Thirdly, Henry argues that one is only commanded to do an act that he can cause. One is not commanded to do what has to be done by another. But the will's activity of willing extends only to the exercise of actions of other potencies, including those of the intellect, or to those actions that the intellect determines for it. Reason does not command the first sort of willing, as even Henry's opponents admit. But they claim that the will does not move itself to the second sort of willing, since that is determined by the intellect. In that case, the will could only be commanded by reason by the sort of rule that Aristotle called despotic. Hence, one should hold that the intellect can in no sense command the will. But since either the intellect or the will must have command, command belongs to the will as the potency that can command the other without qualification.

3. Quodlibet XIV: The Last Word

Henry's fourteenth quodlibetal disputation is dated by Gómez Caffarena as having taken place at Christmas 1290 and by Paulus as having taken place at Easter 1291. In question five, the last selection translated in this volume, we find Henry's final intervention on the subject of human freedom.[64] The question asks whether the intellect and the will

---

62. Aristotle, *Politics* I, c. 5, 1254b4-6.
63. Thomas Aquinas, *The Summa of Theology* (*Summa theologiae*) I-II, q. 17, a. 1 ad 3um.
64. Cf. *Henrici de Gandavo Quodlibet* I, ed. Macken, p. xvii.

are equally free potencies. There are three arguments to the effect that intellect and will are equally free.

The first argument claims that the freedom of a potency lies in its ability to elicit its act first and by itself; that is, a potency is free if its act does not required the previous act of another potency in order to act. Since both the intellect and the will are spiritual potencies, the argument claims that each of them can move themselves to act, and since they are equally spiritual, they are equally free.

The second argument claims that rational potencies can produce contrary effects; that is, a rational potency is not determined to one of two contraries. Since intellect and will are equally rational, they are equally free, or, if one is more rational and, therefore, more free, it is the intellect.

The third argument states that a potency is more free to the extent that it is less dependent upon another potency and its act. But the intellect does not depend upon the will at all, while the act of the will depends upon the intellect's knowing some good. Henry argues in the argument to the contrary that the potency that freely determines itself to its act is more free than the potency that is determined to its act by something else and that the will freely determines itself to its act, while the intellect is determined by the object.

In his answer Henry develops a definition of freedom, beginning with the genus and adding six differences that further specify the meaning of freedom. The genus of freedom is faculty, that is, an ability to act or to be acted upon. The first difference is that freedom is an ability to act, since freedom is not found in passive potencies. The second difference is that freedom is an ability to do or attain what is good. The third difference is that freedom is the ability only to do or attain what is good. Thus freedom of the will does not consist in the ability to will or not to will the good, but only in the ability to will the good. The fourth difference adds that one is free if one can do what is good for oneself without needing another, while the fifth difference adds that one can do this from a principle within oneself. All of these differences still do not differentiate the intellect and the will from natural potencies. That is, all the characteristics of freedom mentioned so far are found in heavy things that move of themselves to their natural place. Hence, Henry adds the further difference, namely, that the potency can do this spontaneously; that is, the sixth difference excludes the impulse by which a natural potency attains its good. Both intellect and will share the genus and these six differences. Hence, if the will's freedom had nothing more, the intellect and the will would be equally free. Henry then points to a seventh difference that characterizes the freedom of the will: the will can restrain the intellect from its act and compel it to act, whereas, the will cannot be compelled to act or restrained from acting by anything. Henry admits that "the intellect is free in some sense," but "the will is much freer than the intellect."[65]

At this point Henry mentions the position of some who hold that the will can begin to act without any compulsion, but cannot cease from acting, since the will necessarily follows the judgment of reason and owes its freedom to the intellect. Thus the intellect would be freer than the will. Macken identifies this position as that of Giles of Rome.[66] Thus Henry comes to the full definition of freedom which belongs to the will alone: "the faculty by which it is able to proceed to its act by which it acquires its good from a spontaneous principle in itself and without any impulse or interference from anything else."[67] Furthermore, even though the intellect has some freedom, it does not have freedom of choice which belongs to the will alone.

Henry then turns to the objections. With regard to the first, he denies that the intellect and will elicit their acts with the same degree of firstness and of themselves, though he grants the claim that the ability to move itself belongs to a potency by reason of its spirituality. He points out that one cannot legitimately infer from the fact that self-motion is impossible only in bodies and bodily powers that self-motion is equally possible for all powers that are not in a body. After all, he points out, angels move themselves more freely than human beings in both intellect and will.

Henry then turns to a lengthy discussion of the text from Averroes where the Commentator had indicated that self-motion was only impossible in bodies and in powers in bodies. Henry aims to show that Averroes did not mean that such self-motion was possible only in separate substances. Once again, Henry's concern with grounding his position solidly in the Aristotelian tradition becomes apparent.

Following Averroes, Henry also points out that motion is predicated equivocally not only of God and of bodies, but also of spiritual potencies, such as the intellect and will, and bodies, though the equivocity is greater between God and a creature than between any two creatures, for example, between the will and a body.

With regard to the second objection, Henry states that freedom is a characteristic of rational potencies that are active and that Aristotle was speaking of active rational powers when he said that rational powers can produce contrary effects. Though the intellect is a rational potency, it cannot act unless it is first acted upon; furthermore, the intellect is not

65. Cf. *Quodlibet* XIV, q. 5, p. 80. In "Heinrich von Gent im Gespräch" (p. 181), Macken mentions Lottin's view that Henry is here attacking Godfrey of Fontaines who had ascribed freedom to the intellect. Following San Cristóbal-Sebastián, Macken suggests that Henry is perhaps alluding to certain ideas of Giles of Rome and is responding to Giles's *Quodlibet* V, q. 15.
66. Cf. R. Macken, "Heinrich von Gent im Gespräch," p. 180.
67. Cf. *Quodlibet* XIV, q. 5, p. 81.

even active in judgment unless the will consents. Hence, though the intellect is just as rational as the will or even more rational than the will, the intellect, as a passive potency, cannot freely produce contrary effects.

In replying to the third objection, Henry admits that the will requires a previous act of the intellect if it is to be determined to its act, but he insists that the intellect is much more dependent upon its object than the will is dependent upon the intellect, because the intellect not merely requires the presence of its object, but also has to be acted upon by it before it can elicit its act. Yet, the will does not have to be acted upon by the intellect. Rather, the known good is merely a condition of the will's acting.

IV. Henry's Influence

Even during his life, Henry met with opposition from men, such as Giles of Rome and Godfrey of Fontaines. After Henry's death the opposition grew among the Dominicans, especially those who followed St. Thomas. Henry's influence was most noticeable upon Duns Scotus, though the latter more often than not differed from Henry's thought. During the fourteenth century a number of Franciscans, particularly at Oxford, showed a preference for Henry's thought. Henry's teachings decisively influenced the Albertist schools in Paris and Cologne during the fifteenth century.[68] In the beginning of the sixteenth century the Servites abandoned Thomas Aquinas and Duns Scotus and adopted Henry as their official doctor, naming their house of studies in Rome after Henry and publishing a new edition of each of his two main works. In the late nineteenth century German scholars stripped away the many items of legend that had accrued to Henry's biography.[69] At the turn of the century, Maurice de Wulf pioneered the modern study of Henry's philosophy.[70] The publication of Henry's *Opera Omnia* in a critical edition that will eventually come to forty-six volumes marks the beginning of a new scholarly approach to the thought of Henry that has already produced many significant studies of the Solemn Doctor and will lead to many more in the years to come. On the topic of human freedom, the articles by Raymond Macken have helped the student of medieval philosophy to

---

68. J. Paulus, *Henri de Gand*, p. xx.
69. Cf. especially F. Ehrle, "Beiträge zu den Biographien berühmter Scholastiker: Heinrich von Gent," *Archiv für Literatur- und Kirchengeschichte des Mittelalters* 1 (1885), 365-401 and 507-508.
70. Cf. his *Histoire de la philosophie scolastique dans les Pays-Bas et la principauté de Liége* (1894-1895). The pages dealing with Henry have been published as *Etudes sur Henri de Gand* (Bruxelles, 1895). De Wulf's *Histoire de la philosophie en Belgique* (Bruxelles, 1910) treats of Henry on pp. 80-116.

examine Henry's thought for its own sake and not merely in contrast with the position of Aquinas and Scotus and to see in his views of human freedom a position much more balanced than the extreme voluntarism that has traditionally been ascribed to him.

Quodlibetal Questions

On Free Will

Quodlibet I, Question 14

Is the will a higher power than the intellect, or the intellect a higher power than the will?

There follows a treatment of questions that pertain both to the separated soul and to the soul joined to the body. One of these concerned the comparison of its two principal powers to each other, namely, whether the will is a higher power than the intellect or the intellect is a higher power than the will. The other five were concerned with the comparison of their actions.

With regard to the first question, it was argued that the intellect would be a higher power, because the Philosopher says this in the tenth book of the *Ethics*.[1] According to him, practical reason is the first mover in things to be done by the will.[2] Moreover, Augustine says in chapter twenty two of *Against Faustus*: Without a doubt in actions of the soul, contemplation, which belongs to the intellect, is preeminent.[3] Moreover, in his reason man is formed anew according to the image of God.[4] Finally, that which directs is higher than that which it directs and the judgment of the intellect directs the will.

Against this view is the fact that the will is the first mover of itself and other things in the whole kingdom of the soul, and such a power is higher.[5]

<The Solution>

To this we must say that, since the powers of the soul of themselves are hidden from us and unknown to us, just as the substance of the soul is, we have to seek, in a way appropriate to us, all knowledge concerning them from what is subsequent to them. Hence, we have to judge the preeminence of one power over another from those things that are subsequent to the powers and that provide us a way of coming to know the powers. These are three: habit, act, and object. We must say that the power whose habit, act, and object are superior to the habit, act and object of another is without qualification superior to that other power.

---

1. Cf. Aristotle, *Nicomachean Ethics* X, 7, 1177b30-1178a2.
2. Cf. Aristotle, *Nicomachean Ethics* I, 13, 1102b28.
3. Cf. Augustine, *Against Faustus* (*Contra Faustum*) XXII, 27: *CSEL* XXV, 621. Where the citation is not exact, as in the present case, I have omitted quotation marks.
4. Cf. 2 Co 3:18 and Col 3:10.
5. Early in the thirteenth century, William of Auvergne had drawn an extended comparison of the will to the king or emperor over the other powers of the soul; cf. his *De anima* c. II, pt. 15, in *Opera omnia* (Orléans-Paris, 1674), vol. II, pp. 85f. See my "The Will as King over the Powers of the Soul: Uses and Sources of an Image in the Thirteenth Century," *Vivarium*, forthcoming.

Now it is the case that the habit, act, and object of the will are utterly superior to the act, habit, and object of the intellect. Hence, we must say that the will is absolutely superior to the intellect and is a higher power than it.

The position we have taken is clear because the characteristic habit of the will which carries it toward the good by an act of true love is the habit of charity. By it, according to Augustine, we love God in himself and the neighbor in God and because of God.[6] But the highest habit of the intellect is wisdom by which we contemplate God and things eternal, according to Augustine in book fourteen of *The Trinity*.[7] The Apostle states well the degree by which the habit of charity is superior to every habit of wisdom and knowledge, when he says in chapter thirteen of the First Letter to the Corinthians, "If I speak with the tongues of men and of angels and do not have charity," and so on.[8]

The degree by which the act of the will, which is to will or to love, surpasses the act of the intellect, which is to know or to have knowledge, is obvious from two comparisons: first, from the comparison of one act to the other, second, from the comparison of each of them in terms of how the subject of the act is perfected by its object.

What we are aiming at is clear from the first comparison. For, as Augustine says in book twelve of *On Genesis*, and the Philosopher says in book three of *The Soul*, "The agent and the mover are always more noble than that upon which they act."[9] But the will is the universal and first mover in the whole kingdom of the soul and superior to and first mover of all other things to their end, as will be seen below. For, as Anselm says in *Likenesses*, "It moves reason and all the powers of the soul."[10] And as Augustine says in book three of *Free Choice*, "The mind itself is first subject to the intention of the mind; then the body which it governs, and thus it moves any member to activity."[11] Hence, the will commands reason to consider, to reason, and to deliberate when it wills and about what topics it wills, and it likewise makes it to stop. The intellect does not command or move the will in any such way, as will become clear further on, when we say more about their comparison.

From the second comparison, what we are aiming at is likewise clear. For by the action of the will the will itself is perfected by the very reality that is loved as it exists in itself, because by its action the will is

---

6. Cf. Augustine, *On Christian Doctrine* (*De doctrina christiana*) I, xxii, 21: *CC* XXXII, 17-18.
7. Cf. Augustine, *On the Trinity* (*De trinitate*) XIV, i, 3ff.: *CC* L/A, 422ff.
8. 1 Co 13:1ff.
9. Cf. Augustine, *The Literal Interpretation of Genesis* (*De Genesi ad litteram*) XII,16: *CSEL* XXVIII, 402, and Aristotle, *On the Soul* (*De anima*) III, 5, 430a18-19.
10. Pseudo-Anselm, *Likenesses* (*De similitudinibus*) 2: *PL* CLIX, 605C.
11. Augustine, *Free Choice* (*De libero arbitrio*) III, xxv, 75: *CC* XXIX, 320.

inclined toward the reality itself. But by the action of intellect the intellect is perfected by the thing known as it exists in the intellect. By its action the intellect draws into itself the reality known, while by its action the will transfers itself to the object willed for its own sake so that it may enjoy it. For this reason, as Dionysius says in chapter four of *The Divine Names*, by its action the intellect likens itself to the reality known, but the will transforms itself into the object willed.[12] It is much more perfect and lofty to be transformed into the good as it is in itself according to its own nature than to be made like the true as it is in the knower in the manner of the knower and thus in an inferior manner. Accordingly, Augustine says in the eleventh book of *The Trinity*, "When we know God, his likeness comes to be in us, but a likeness of an inferior degree, because it is in an inferior nature."[13] Hence, the activity of the will is far more perfect and lofty than the activity of the intellect to the degree that love and esteem for God is better than knowledge of God. Even if with respect to those things that are less than the soul the opposite is the case, namely, that the action of the intellect is higher than the will, because the knowledge of bodily things in the soul is higher and more noble than the love of them, this only makes the intellect to be more noble than the will in a certain respect. But the first relation and comparison makes the will to be higher without qualification. For the first goodness and the first truth are the essential and primary objects of the intellect and the will; other things are objects of the intellect and the will in comparison to them secondarily and in a certain respect. In the same way, in other things something true or good is true or good in some respect in comparison to the first truth and first goodness, since by nature it does not have the character of true or good except through an impression of the first truth and goodness, as will have to be explained elsewhere. Thus the will seeks something good by reason of some participation that thing has in the first goodness and the intellect knows something true only by reason of some participation that thing has in the first truth. Accordingly, it is more natural for the will to be perfected by the first goodness than by anything else and for the intellect to be perfected by the first truth than by anything else. For this reason the will and the intellect cannot perfectly come to rest in the enjoyment of any good or in the knowledge of any truth until the first goodness and the first truth are attained. In accord with this, Augustine says in the beginning of *The Confessions*, "You have made us for yourself, and our heart is restless until it rests in you."[14] Hence, since everything should be judged to be unqualifiedly more of a certain kind in comparison to that

---

12. Cf. Pseudo-Dionysius, *The Divine Names* (*De divinis nominibus*) IV, #4: *PG* III, 711C-D.
13. Augustine, *The Trinity* (*De trinitate*) IX, xi, 16: *CC* L, 307.
14. Augustine, *The Confessions* (*Confessiones*) I, i, 1: *CC* XXVII, 1.

which is more without qualification and more in terms of its nature, as the Philosopher says in the first book of *Posterior Analytics*,[15] the act of the will should be judged unqualifiedly better than the intellect and absolutely so, since it is unqualifiedly better than it in comparison to its first object. This agrees with the thought of the Philosopher in the *Topics*: "If the best in this genus is better than the best in that genus, then the former is better than the latter without qualification."[16]

Next, that the object of the will is superior to the object of the intellect is obvious, because the object of the will, which is the good without qualification, has the character of an end without qualification and of the ultimate end. The object of the intellect, which is the true, has the character of a good of something, for example, of the intellect. Thus it has the character of an end subordinate to another end and ordered to the other end as to the ultimate end. For, when there are many particular ends, they are all included under some one end, and all the powers which have divers ends are subordinated to some one power whose end is the ultimate one, as is stated in the beginning of the *Ethics*.[17] In accord with this, then, the intellect is completely subordinated to the will. And in this way, as in all active potencies ordered to an end, that potency which regards the universal end always moves and impels to activity the other potencies which regard particular ends and regulates them, as the master art regulates the other arts in a city, as is stated in the beginning of the *Ethics*,[18] so the will moves the reason and directs it to activity, as well as all the powers of the soul and members of the body.

It must, then, be said that the will is absolutely the higher power in the whole kingdom of the soul and thus higher than the intellect.

<With Regard to the Arguments>

It is easy to reply to the objections raised against this position.

To the first objection, with regard to what the Philosopher says in book six of the *Ethics*, one should say that his comparison is literally understood with regard to those potencies in which there are the other intellectual habits, and thus nothing from that statement applies to the will.

To the second objection, that practical reason is what moves first, one should say that something is said to move in two senses. In one way, metaphorically, by proposing and revealing an end toward which one should move. Practical reason moves in this way, and in this way it moves

---

15. Cf. Aristotle, *Posterior Analytics* I, 2, 72a29-30.
16. Aristotle, *Topics* III, 2, 52c.
17. Cf. Aristotle, *Nicomachean Ethics* I, 1, 1094a6-10, 18-19.
18. Cf. Aristotle, *Nicomachean Ethics* I, 1, 1094a4-5, 9-10.

the person who wills; it does not, properly speaking, move the will, which is moved by the person who wills. Nor does reason, properly speaking, move in this sense; rather, it is the object that of itself moves reason to know and, thereby, in revealing itself as good, it metaphorically moves the person who wills to desire it. For the good as known moves the person who wills, but reason itself as knowing does not move the will. In another way, something is said to move another in the manner of an agent and one impelling the other to act. In this way the will moves the reason, and this is more truly to move.

To the third objection that contemplation holds the first rank in the actions of the soul, one should admit that it is true, but this has nothing to do with the will, since he was speaking about the relation of the active and the contemplative life. Of these the one is ruled by speculative reason which is the higher; the other is ruled by practical reason which is lower. But both are ruled by the will which is above both of them.

To the fourth objection, that the image is formed anew in reason, one should say that it is true, but not the whole truth. For part of the image, and the perfecting part, pertains to the will. For this reason the mental word in which the perfect character of the image shines forth, is, according to Augustine, "knowledge along with love."[19]

To the fifth objection, that what directs is superior to what it directs, one should say that there is one who directs with authority, as a lord directs a servant; he is the higher. In that way the will directs the intellect. Or, one directs another by way of service, as a servant directs a master in carrying a light before him at night so that the master does not stumble. Such a director is inferior, and in this way the intellect directs the will. Hence, the will can withdraw the intellect from directing and knowing when it wills, as a master can withdraw a servant.

Quodlibet I, Question 15

Does the act of the will precede the act of the intellect, or the act of the intellect that of the will?

After the question on the comparison of the intellect and the will in terms of their essence, there followed five questions concerning their comparison in terms of their actions.

First, does the act of the will precede the act of the intellect, or the act of the intellect that of the will?

---

19. Augustine, *The Trinity* (*De trinitate*) IX, x, 15: *CC* L, 307.

Second, when reason has proposed something good and something better, can the will chose the lesser good?

Third, is the disorder of the will caused by an error of reason, or is an error of reason caused by the disorder of the will?

Fourth, is the will evil if it disagrees with reason when reason is in error?

Fifth, does the will sin more by acting against reason when reason is in error than by acting in accord with reason when reason is in error?

With regard to the first question it was argued that the act of the intellect precedes the act of the will. For agents that act by nature without any knowledge of their own must have another knowing being superior to them by which they are directed to an end grasped by that knower before they move. Hence, when the knower and the mover are joined together in the same being, as are the intellect and the will, the knower must know before the other can move.

On the contrary, the first principles of the sciences are first sought because of the desire to know. But desire is an act of the will. Hence, it precedes the knowledge of the principles. And thus it precedes the knowledge of everything else.

<The Solution>

To this question one should say that, if we are speaking indiscriminately about any knowledge whatsoever, it is clearly impossible that the will will something without some previous knowledge. For this reason, Augustine says, we can love what is unseen, but never what is unknown.[1] After all, that of which we have absolutely no knowledge – neither generic nor specific, neither by the senses nor by the intellect – can in no way be an object for the will. Thus some sort of knowledge must always precede the will. The will, of course, can precede further knowledge, since imperfect knowledge kindles the desire for perfect knowledge, as Augustine stated.[2]

But since there are in human beings two kinds of knowledge, the one sensitive and the other intellective, the question arises whether one can be moved to will something if only sensitive knowledge precedes without any intellective knowledge. For Augustine says in book three of *Free Choice* that the human will can be stirred to will by each kind of knowledge. He says, "One must admit that the mind is stirred by higher and lower things that it sees so that the rational substance takes what it wills from each of these. One who wills certainly wills something. But unless

1. Cf. Augustine, *The Trinity* (*De trinitate*) VIII, iv, 6-9: *CC* L, 274-284.
2. Cf. Augustine, *The Trinity* (*De trinitate*) IX, xii, 18: *CC* L, 309-310.

one is moved externally by the senses of the body or something comes to mind in hidden ways, one cannot will."[3] For this reason the question arises whether, if some interior organ is destroyed which in turn destroys human knowing, as the Philosopher says, one might will something through the knowledge derived from those things known by the senses, since he still has sound senses by which he perceives many differences in things.[4] Or is all appetition regarding them sensitive and merely animal?

One should say that in those things which are not moved by natural inclination alone, an appetite not only requires knowledge, but a particular appetite also requires a particular knowledge. Thus, if the sensitive appetites were distinguished in terms of the distinction of the senses, if a particular sense were lacking, the appetite corresponding to that sense would have to be lacking. In the present case the rational appetite corresponds to rational knowledge, just as some particular appetite regarding colors would correspond to the sense of sight.

One should state absolutely that knowledge of the intellect must precede the action of the will. Without such previous knowledge it can will nothing. As a result, in insane persons whose intellects are destroyed, there is no appetite of the will, but only the sensitive appetite of an animal. For, if the intellect is taken away, the human being remains only as an animal. In accord with this the Philosopher says at the end of book three of the *Politics*, "He who commands the intellect to rule seems to command God and the laws to rule, but he who commands a human to rule appoints an animal as well."[5]

<With Regard to the Arguments>

The answer to the opposing argument is clear from what has been said. For a human being does not proceed to the investigation of principles from a desire to know without some general knowledge by which the will is stirred to know something in particular. In this way philosophers, seeing the effects, but not knowing the causes, began to philosophize and seek out the causes of things, as is said in book one of the *Metaphysics*.[6] For they knew in general that effects did not occur without a cause, for instance, an eclipse of the sun and moon.

---

3. Augustine, *Free Choice (De libero arbitrio)* III, xxv, 74: CC XXIX, 319.
4. Cf. Aristotle, *Nicomachean Ethics* VII, 3, 1150a25.
5. Aristotle, *Politics* III, 16, 1287a27-30.
6. Cf. Aristotle, *Metaphysics* I, 2, 982b12-21.

Quodlibet IX, Question 5

Does the will move itself?

Regarding the second point it is argued that the will does not move itself. Since it is simple and one and the same thing, the same thing would then be with respect to the same thing both in act and in potency at the same time, because a mover, insofar as it is a mover, is in act with respect to the moved, and the moved is in potency with respect to the mover. But that is impossible, because act and potency with regard to the same thing are contraries and are not compatible with each other.

On the contrary, a heavy thing, which is something material, moves itself, when the cause of the heavy thing ceases and there is no impediment. Likewise, animals move themselves in forward motion. Hence, for even better reasons, the will, which is something immaterial, moves itself.

<The Solution>

Here one must consider six kinds of things which stand in an order in terms of what it means to move. In accord with their order they have a greater or lesser difference between the mover and the moved, for in the first level, the mover and the moved differ less than in the second, and so on. This is the case when one takes motion in the widest sense to include motion that is real and most properly so called and motion that is only in the mind and most improperly so called.

It is universally true in all things that move and are moved that the power and meaning of the mover and the moved are necessarily contraries. For, as the first objection urges, the power and meaning of the mover comes from the state of perfection by which it is in act and influences something else. The power of the moved, on the other hand, comes from the state of imperfection by which it is in potency and receives something from something else for its perfection. Thus something that moves in every respect without being moved, such as God, and something moved in every respect without moving in any sense, such as primary matter, are most distant from each other in their essence. For God is pure act and supremely perfect, while matter is pure potency and supremely imperfect.

Those six levels of things which stand in an order, as we said, in terms of what it means to move are the following: The first is the divine will in willing; the second is the created will in willing. The present question concerns this latter will, and on its account we introduce a discussion of the other things for the sake of a better explanation. The third is the created intellect in understanding; the fourth is a heavy or a light thing in moving itself. The second objection is based on this. The fifth is the animal in moving itself by forward motion. The sixth is the cause in moving something separate from itself to produce something else.

The first of these moves itself in willing or to the act of willing, and the will as moving and as moved differs only in our minds, just as the volition differs from these only in our minds.[1] Hence, by speaking in the widest sense and by an extension of the name, God's willing, or act of willing, is called, as we think of it, motion or a movement in the divine will, and the will itself, insofar as it is moving, is, as we think of it, like something perfect and in act, while the volition is like the motion or action of that which exists as in potency. After all, it is impossible to say that, insofar as it is in potency, it moves itself. Only insofar as it is in act, does it move itself to the extent that it is in potency.[2]

But the other things that move and are moved, to the extent that they are more remote from God by the order of nature, have a greater and lesser difference between the mover and the moved. Accordingly, the last mover and moved, the mover and moved at the sixth level, differ from each other in substance and essence and are separate in location and mass, while the things that move and are moved in the other intermediate levels do not differ completely in substance or location.

Hence, in the fifth level at which the animal moves itself by forward motion, the mover and the moved are partly different in substance and partly not, and they are partly separate in location and partly not.

And to begin from the higher, the Philosopher determined in the third book of *The Soul* that the first "mover in terms of place is not the vegetative power," because there are no organs of forward motion corresponding to it in "plants. Nor is the speculative intellective power" the first mover, because things to be done – for the sake of which motion exists – are not matters of speculative knowledge.[3] Nor is the practical intelligence the first mover in terms of what it determines. In accord with this, Themistius comments on this point more precisely, "The practical intellect is not the master of motion," because "it often considers something worthy of flight, say, an earthquake or a wild beast, and the heart leaps; yet the animal remains in place. So too, a part of the body perceives something pleasing; yet the whole animal remains at rest, as though there were some other master of its movement. But neither desire nor anger is master of motion; after all, men of self-control, though angry and full of

---

1. That is, there is a purely mental or rational distinction between the mover and the moved in God.

2. The Aristotelian definition of motion is: the act of a being in potency insofar as it is in potency. Cf. *Physics* III, 1, 201a10-12. It is impossible, in accord with that definition, for something to move itself. Hence, Henry takes motion in the widest sense, extending the meaning of motion even to God who is utterly without any real motion.

3. Cf. Aristotle, *The Soul* (*De anima*) III, 9, 432b14-27.

desire, become quiet." And in conclusion he states that "these two move: the appetite and the practical intellect,"[4] but the intellect only moves because the appetite moves. Hence, in the forward motion of animals, the soul of the brute animal is the mover in accord with the sensible appetite, but the human animal, insofar as it is human, is moved in accord with the rational appetite, which is called the will. But the whole composite is moved.

Since an animal is moved only by its organs, there must be some organ which the soul uses for motion in accord with its appetite, and in that appetite we must assign what the animals seeks as the mover. Hence, Averroes says, "A body is moved by the first organ so that the first organ which moves and which is the subject of the desiderative soul is in one place in the body of the animal, from which" certain "parts are pushed and to which other parts are pulled."[5] Along with this, there must be in it some one thing which can be the beginning and the end of motion. "For in all motion produced by pulling and pushing the beginning from which the pushing is made must be the end to which the pulling draws back, as in a circular motion. But since the motion of an animal is produced by pulling and pushing, it is clear that, when the right leg is moved by us and we are supported by the left, then certain parts of that leg are pushed forward, as the front parts, and others are pulled back, as the back ones. And the pulling and pushing of them are not in a straight line, but along curved rather than straight lines, and for this reason it is like a circle."[6] This part at rest is the end by reason of the motion of pulling and the beginning by reason of the motion of pushing. Thus there must be a socket there "so that the midpoint is at rest in the socket from which the pushing will begin and at which the pulling will end, and so that the motion will be curved both on the right and on the left. In this motion the beginning and the end as the pivotal point are different in their definition, but the same in their mass. In the animal it is the heart that is such."[7]

Averroes had not seen what Aristotle said in the books, *The Forward Motion of Animals* and *The Causes of the Motion of Animals*, since he did not have them in translation, as he says in this passage.[8] But for a better understanding of this doctrine, one should know that, in every movement of an animal by itself, it is a general truth that the moved

---

4. Themistius, *The Soul* (*De anima*) VII, tr. Moerbeke, ed. G. Verbeke, p. 263, ll. 4-9.

5. Averroes, *The Commentary on the Soul* (*De anima III Comm.* 55), ed. Crawford, p. 525, ll. 16-20.

6. Averroes, *The Commentary on the Soul* (*De anima III Comm.* 55), ed. Crawford, p. 525, l. 20-p. 526, l. 33).

7. Averroes, *The Commentary on the Soul* (*De anima III Comm.* 55), ed. Crawford, p. 526, ll. 55-59.

8. Cf. Averroes, *The Commentary on the Soul* (*De anima III Comm.* 54), ed. Crawford, p. 524, ll. 59-61.

mover has of itself some fixed and immobile point on which it rests and is supported, and it has such a fixed point both within itself and external to itself.

Regarding its needing such a fixed point within it, the Philosopher says in the beginning of *The Causes of the Motion of Animals*, "In all motion of animals, if some part is moved, another part must be at rest, and this is why animals have joints. For they use these joints as pivotal points, and the whole member in which there is the joint becomes both one and two, both straight and bent, with changes in potency and act on account of the joint, and those things which are in potency and in act in the joints are at times one, but at other times divided. But the first source, insofar as it is the source, is always at rest, while the part below it is moved. Thus, for example, the elbow is at rest when the forearm is moved, the shoulder is at rest when the whole arm is moved, the knee is at rest when the shin is moved, the hip is at rest when the whole leg is moved."[9] And later he says, "Hence, if the animal were the forearm, the principle of the moving soul would, of course, be here."[10] And below he says, "The principle of the moving soul must always be in the middle of both. For the middle is the last term of both extremes."[11]

But regarding the fact that there must be such a fixed point outside the animal, he says in the same work between the passages already mentioned, "It is obvious that each individual must have something at rest within itself from which a motion begins. But everything else which is in the individual is insufficient, if there is not something external which is simply at rest and immobile. For just as in the individual there must be something immobile, if it is going to move, all the more so, there must be something immobile outside the animal, and by pushing against that, what is moved moves. For if it were always giving way, as, for example, when mice walk in the sand, the animal makes no progress unless the earth stands still, nor would there be flying or swimming if the air or the sea did not resist."[12] He says in *The Forward Motion of Animals*, "That which is moved always changes place, supported by what is beneath it. For this reason, if what is under foot either is carried off faster than what is moving in it can find support or is carried off entirely, and if it does not offer any resistance at all to those things that move, nothing can move in it."[13] As he says in *The Causes of the Motion of Animals*, "That which offers resistance must be other than that which is moved, and the whole of it must be other than the whole of what is moved, with no part

9. Aristotle, *The Motion of Animals* (*De motu animalium*) c. 1, 698a15-b4.
10. Aristotle, *The Motion of Animals* (*De motu animalium*) c. 8, 702a31-32.
11. Aristotle, *The Motion of Animals* (*De motu animalium*) c. 9, 702b15-17.
12. Aristotle, *The Motion of Animals* (*De motu animalium* c. 1, 698b4-7, 12-18.
13. Aristotle, *The Forward Motion of Animals* (*De animalium incessu*) c. 3, 705a7-12.
14. Aristotle, *The Motion of Animals* (*De motu animalium*) c. 2, 698b18-20.

belonging to that which is moved."[14] Further on he says, "All these things must have in themselves that which is at rest and something external against which they are pushed forward."[15] "Hence," as he says in the same place, "these are necessary not only in those things that move of themselves in place, but also in those things moved in inhalation or exhalation," and other movements by themselves.[16]

For a specific knowledge of the manner of the forward motion of animals that is touched upon in the argument, he says in *The Forward Motion of Animals*, "That which is moved always uses two organic parts by which it produces the motion, and it does this by distinct parts in the same animal on the right and on the left."[17] In accord with this, he adds after a bit: "One part of the body is by nature the beginning of change according to place. This, of course, is the right, but the opposite whose nature is to follow is the left."[18] As he says in *The Motion of Animals*, "In the midst of both there must be the principle of the moving soul, but what moves both must be one. This is the soul, which is something other than such a mass, in some animals in the heart, in others in the analogue of the heart."[19]

According to what he determines in the third book of *The Soul*, it is necessary to consider here four items that are found in an order.[20]

The first of these is the unmoved mover which is the known desirable good.

The second is the moved mover, but it is moved by something other than what it moves. This is the appetitive power of the soul moved by the known desirable good by a movement involving a twofold change, and this happens in both the rational and the sensitive appetite. For the good to be sought and the evil to be avoided cause sensations, imagery, and ideas by producing changes. For images and ideas have the power of real things and bring about modifications in the appetitive potencies. Images, first of all, change the sensitive appetitive powers in their organs in terms of hot and cold, and immediately thereafter they change them by motion of the appetite so that it begins actually to desire after not desiring. When the sensitive appetite has been moved in this way by the movement involving the two changes, it immediately moves by impulse the organic member by the twofold local motion of pulling and pushing, as will now be seen, in the way it moves other things by a like movement. For, according to what the Philosopher says toward the end of *The Motion of Animals*, "When the place around the organ is changed by sen-

---

15. Aristotle, *The Motion of Animals* (*De motu animalium*) c. 4, 700a18-23.
16. Aristotle, *The Motion of Animals* (*De motu animalium*) c. 4, 700a18-23.
17. Aristotle, *The Forward Motion of Animals* (*De animalium incessu*) c. 3, 705a19-21.
18. Aristotle, *The Forward Motion of Animals* (*De animalium incessu* c. 4, 705b18-21.
19. Aristotle, *The Motion of Animals* (*De motu animalium*) c. 9, 702b16-17 and 703a1-3.
20. Cf. Aristotle, *The Soul* (*De anima*) III, 10, 433b13-19.

sation or changed in terms of warmth or coldness or in terms of some
other such modification, there occurs a change around the heart, and the
parts connected with it change along with it, expanded and contracted so
that motion necessarily comes about in animals on this account. And it
makes a big difference in the body in blushing, blanching, in fears and
tremors and their opposites."[21] But when something to be done is known,
the intellect changes the will by a certain modification involving a spiritual
change, as we will see further on. Upon this there follows a change in
appetite when it begins to will after not willing. And then by its command
it moves the organs by local motion, just as the sensitive appetite moves
them, as we have said.

Hence, the third moved mover – and moved by the same kind of
motion as that by which it moves, namely, local motion – is also an organ
of the appetitive and moving power in the heart or in its analogue. This,
as we have already said, is moved by the appetitive power of the soul by
pulling and pushing and, when moved, it moves the other connected parts
and, consequently, those joined to them. But such a motion is found
successively, now in the right part, now in the left, and not simultaneously
in both, because, as one part is moved,[22] the other must be at rest, as has
been said and will now be explained more fully.

The fourth is moved and not moving, and these are the connected parts and external members of the animal. Nonetheless, that organ
is moved according to its parts, namely, right and left, by the appetitive
power in pulling and pushing in other ways than the connected parts are
moved by the same motion, as will now be stated.

At the end of the third book of *The Soul*, the Philosopher
reduces the four just mentioned to three. He says, "The first mover
pushes and is not pushed; the last mover, however, is pushed and does not
push. But the middle one both pushes and is pushed, and there can be
numerous middle ones."[23] Thus he reduces what is mover and moved to
one. Still it is of great value for our purpose to see their distinction
according to another manner of moving, as will now become clear. Thus,
as the Commentator says on this passage, "motion in these things has
three components: the mover, the middle element, and the last thing
moved. But the middle element can be one thing or more than one."[24]

That organ, which we have already discussed, surrounds the principle which is the middle of the heart or its analogue. In it the appetitive
and moving power of the soul principally resides, as in an indivisible point

---

21. Aristotle, *The Motion of Animals* (*De motu animalium*) c. 9, 702b21-25 and c. 7, 701b28-29.

22. Though the critical edition has "*nota*," I have translated "*mota*."

23. Aristotle, *The Soul* (*De anima*) III, 12, 434b32-35.

24. Averroes, *The Commentary on the Soul* (*De anima III Comm*. 65), ed. Crawford, p. 539, ll.31-33.

which is surrounded by the organ with its distinct parts on the right and
the left, front and rear, top and bottom, though they are principally on the
right and the left. The whole organ is one in power, but two in operation,
because while one part, for example, the left, is at rest, the other is
moved, for example, the right, but this is done in turns. All these points
are gathered together from where they are scattered in *The Motion of
Animals*. He omitted the explanation of these things in *The Soul*,
promising in it an explanation of them in this work.[25] And though, as was
said, when one of the parts we mentioned is at rest, the other is in its turn
moved, neither of them is, nonetheless, moved by the other. Rather both
are moved by an immobile third part, which is the soul. For, as is said in
the eighth book of the *Physics*, "it never happens that it moves itself so
that each of the two parts is moved by the other" so that there is
reciprocal motion.[26] As he says toward the end of *The Motion of Animals*,
speaking of the right and the left, "In the organ mentioned, they are
moved by contraries so that it is not the case that the left is moved by
resting upon the right, nor that the right is moved by resting on the left.
Rather, the principle of motion is always in something superior to both,
and it is necessary that the principle of the moving soul be in the middle.
For the middle is the last term of both extremes."[27] As this middle part is
the principle of motion between the right and the left, so it is likewise
among the parts according to other differences of position. Hence, he
immediately adds, "But this holds true in a like manner for other motions,
both for those from above and from below, for example, those which
come from the head, and for those which come from the spine."[28] In no
way, then, is there reciprocal motion existing in this case, because it is the
common term of both, except in the sense that, as was said, that middle
part is the term of both, existing in both and moving both, but the right as
it rests on the left and the left as it rests on the right, as will now be
explained. On this account as well, the right can be said to be moved by
the left, and vice versa. Hence, because these parts have to be moved and
brought to rest by the soul one after the other and through those external
parts of the animal, he says toward the end of *The Motion of Animals*
"that it is necessary that this organ not be a point, but a certain
magnitude,"[29] and that it be divided into the right and the left.

   In accord with this let us understand these three things in every
motion by which an animal moves of itself.

---

25. Aristotle, *The Soul* (*De anima*) III, c. 9, 432b11ff.
26. Aristotle, *Physics* VIII, c. 5, 257b13-15.
27. Aristotle, *The Motion of Animals* (*De motu animalium*) c. 9, 702b13-17.
28. Aristotle, *The Motion of Animals* (*De motu animalium*) c. 9, 702b17-19.
29. Aristotle, *The Motion of Animals* (*De motu animalium*) c. 9, 702b31.

First, there is one indivisible point in the middle of the organ mentioned between the six different positions, namely, right and left, front and back, top and bottom, although it is chiefly between right and left by which that organ is principally distinguished. Thus the Philosopher says in *The Forward Motion of Animals*, "There must be something common by which these things are continuous with each other, and there must be in it the principle of the motion of both parts, as well as of their standing still, in accord with the connection of the parts mentioned. I mean, of the right and left parts, and those which are on the top and the bottom, and which are in the rear or in the front."[30] This point is necessarily at rest, because in all the changes there must be a bending and a straightening, and there must be a beginning and an end so that a slight motion in the sensible part around that middle, whether according to qualitative or local change, produces considerable motion in the other parts of the body according to its size. For, as he says in *The Causes of Motion*, "A slight change made in the beginning produces many big differences as a result. In the same way, when the rudder is moved a slight degree, there is big shift in the prow."[31]

Secondly, we understand the parts of the organ which are near to that point and surround it in continuity with it. Of these right and left are understood to be especially pertinent, and each of these has in itself front and back, top and bottom.

Thirdly, we understand here the remaining connected parts in the animal, both internal and external, that surround that organ and that are moved at the same time as the organ is moved. They are moved much by a small motion in the organ, and their motion is greater to the extent that the parts are more external and further removed from the middle and from the organ. In the same way, in a spherical body moved by a spinning movement, the parts outside on the surface move the most and at the highest speed, while the parts around the center move least and most slowly. It is the same way, if by a single motion wheels are moved at the same time, the smaller wheel being within the larger one, "for the smaller wheel," as the Philosopher says, "is like the center of the larger one."[32]

The motions of the organ and of the connected parts do not differ only in this respect, since in that case there would not be an essential distinction in the motion of the animal between that organ and the other parts so that the organ would be said to be moving and moved, while the other parts are only moved. Rather, they differ in the manner of pulling and pushing, which are the principles of local motion. In the right and left parts, pulling and pushing takes place in two ways. It principally takes place in these parts, but also takes place in the same ways in the others

30. Aristotle, *The Forward Motion of Animals* (*De animalium incessu*) c. 6, 706b18-28.
31. Aristotle, *The Motion of Animals* (*De motu animalium*) c. 7, 701b24-28.
32. Aristotle, *The Motion of Animals* (*De motu animalium*) c. 7, 701b5-6.

parts according to the other differences of position. It takes place in one way in the parts of the organ, in another way in the connected parts.

In one way there is a pulling of a part through its contraction and shortening, as we see in the animal that is called an earthworm. It contracts its front or back parts toward the middle of its body, while it is extended in length, and the contracted part turns red and becomes shorter, when it is contracted or retracted. In another way, there is a pulling of a part without its contraction or shortening, as when someone in bending his knee draws his leg toward himself. As a result the leg is not contracted and does not turn red or become shorter.

Likewise, in one way there is the pushing of a part through its extension and elongation, as we see in an earthworm which first contracts by pulling back the part that it later extends in its forward motion, as if pushing it away from itself. In another way there is a pushing of a part without extending or lengthening it, as when one straightens a bent knee and in that way pushes his leg away from himself. As a result, the leg does not become more extended or longer.

In the first ways, the pulling and pushing occur in the right and left parts of the organ which are close to the principle we mentioned, as the Philosopher says in *The Motion of Animals*, "Hence, this spirit seems naturally suited for there to be motion and to exert strength. The functions of movement are pushing and pulling. Hence, the organ must be able to expand and contract. This is the nature of the spirit,"[33] which the soul uses as an instrument in moving that organ. But pulling and pushing are produced in other ways in the connected parts. Thus, in the parts of the organ, pulling and pushing are produced as in a sentient or living being, namely, with the soul, as it is in one part, pulling toward itself the other, though it principally pulls as it is at rest in the midpoint. For this reason, this organ is a moved mover, not so much because the connected parts are moved by this part, but because by one part another is moved, as happens in the movement of the parts of the earthworm. Hence, he says toward the end of *The Causes of Motion*, "This happens with reason, for we say that it is the power of sensing. Hence, when it is changed because of sensation," and so on, as above.[34]

But in the connected parts pulling and pushing do not occur as they do in a sentient being, but as in something lifeless that contributes nothing to the motion, just as a hand holding a stick, when pulled back, pulls the stick back at the same time. Nor is there a difference insofar as the connected part is something moved, not something moving. Hence, the Philosopher says a little before the statement already cited, "But since it happens that one also has in the hand some lifeless thing, for example,

---

33. Aristotle, *The Motion of Animals* (*De motu animalium*) c. 10, 703a15-22.
34. Aristotle, *The Motion of Animals* (*De motu animalium*) c. 9, 702b20-22.

if one moves a stick, it is clear that the soul will not be in either of the two extremities, neither in the extremity of that which is moved, nor in the other extremity, the principle. For the wood has both a beginning and an end in the hand. Thus, for this reason also, even if the moving principle is not present in the stick, it is present by reason of the soul. Nor is it in the hand; the hand likewise has an extreme at the wrist, and this part at the elbow. It makes no difference whether they are naturally attached or not. The stick becomes like a detachable member."[35]

And so, the principle that moves in terms of place in no sense resides in the connected parts, as the mover, but only as moved. But in the parts of the organ it exists as moved and mover. In the midpoint, however, it is only moving and not moved, save accidentally, insofar as the midpoint has being only in the extremities moved. Hence, the Philosopher says at the end of *The Causes of Motion*, "We should think that the animal is constituted like a city well ruled by laws. For once order has been established in a city, there is no need for a separate ruler who has to be present at every single thing that happens. Rather, each individual does his task, as he has been ordered, and one thing takes place after another by custom. But in animals the same thing comes about by nature. Because each of the constituent parts naturally does its proper task, there is no need that the soul be in each of them. Rather, it exists in something that is the principle of the body, and other things are alive because joined to it and perform their proper function on account of nature."[36]

This principle of the body is chiefly located in the middle of the heart in relation to the whole body, when the whole body is moved from place to place. It is, nonetheless, in some fashion in every joint, when, while the whole body remains in the same place, a part of it is moved in some midpoint between rest and motion in terms of the relation of the moved part to the part at rest. Animals use the joints as pivotal points, as we established above,[37] and pulling and pushing occur in the connected parts, while they retain the same shape and same size. This happens in the same way in the movement of instruments that push and pull one another, though in organic parts the shape and the size are changed, as the Philosopher says later in *The Causes of Motion*, "In puppets and wagons there is no qualitative change; even if the wheels within them become smaller, they still move in a circle. But animals can grow bigger and smaller and can change in shape, when their members have grown larger."[38]

---

35. Aristotle, *The Motion of Animals* (*De motu animalium*) c. 8, 702a32-6.
36. Aristotle, *The Motion of Animals* (*De motu animalium*) c. 10, 703a29-b2.
37. Cf. above, p. 39, para. 4.
38. Aristotle, *The Motion of Animals* (*De motu animalium*) c. 3, 701b10-15.

In fact, although in the connected parts they are not changed in shape or size, it is still necessary that they change in both ways in the parts of the organ. For in them the part that is moved is pulled back toward the midpoint that is at rest, not in a straight line, but in a curved line. This pulling back is produced when the parts, insofar as they are moved by the motion of contraction, are moved from the front and higher toward the rear and lower at the midpoint. The part at rest from forward motion remains in a straight line through the whole motion of the bending of its other part. But that straight line, insofar as it touches the midpoint, leans toward the back at the beginning, although as it begins to bend, it always leans more and more to the front. Thus, both because of the bending of one part and the straightness of the other, the midpoint must stretch toward the rear and, for this reason, make a curvature in the parts so that it is concave toward the front. Thus, if the right and left parts are pulled back at the same time, there is a bending forward of the right and left parts in an arc, while the curvature is toward the back and the midpoint is in it as if in the figure of a semicircle. That is what happens in the motion of jumping.

For in the motion of jumping, at the beginning, the midpoint lies between the right and the left as if between straight lines in a triangular figure. The motion and the bending comes about on the right and the left, and the midpoint is in the curvature in the curved lines. When they are straightened again, they return to the triangular figure. In this pulling back and straightening out, the parts have from one another mutual support and resistance, besides the external support, by which the moving principle, which is in the midpoint, draws back to itself in pulling and drives away from itself in pushing and by which the animal moves as a whole in a jump. Hence, the Philosopher says in the beginning of *The Progression of Animals*, "Something that jumps makes the jump by bringing what is above to bear upon the foundation and what is under its feet. For, in bending, the parts have a certain resistance to one another, and in general what is stationary has a resistance to what presses down on it. For this reason those who carry weights jump further than those who do not, and runners run faster, if they swing their arms in the opposite direction."[39] For there is some resistance produced for the hands and the body.[40] For in the motion of a jump it is as if they move their legs forward, while holding their back the other way.

---

39. I have departed from the critical edition which reads, "runners run faster than those who swing their arms," and followed an alternative reading which seems to make better sense.

40. Aristotle, *The Forward Motion of Animals* (*De animalium incessu*) c. 3, 705a12-19.

But if the right and the left parts of the organ are not pulled back and extended at the same time, but alternately, the one being bent, when the other is straightened out, then, it is extended toward the front in the lower part and toward the back in the higher part. It is joined to the common point in a straight line, and the part pulled back rests upon the other and is united to the midpoint in a curved line. And it is supported in its elevation and bending on the right and finds resistance in it in its extension. This takes place without qualification in the walking motion of those things that take steps. When one part pushes forward, the other takes the weight, and the one raised up passes the burden to the one that supports. Thus they are arranged in a figure on one side having a straight line, but on the other side a curved line. In such forward motion the right part is naturally the source of motion. Thus, it is naturally first to be pulled back rather than the left and to be raised over it by being bent. After being raised up and pulled back, it is extended again, so that motion begins in this way from the right side. The right foot is put forward first, and afterward, when the right foot is at rest, the left is pulled back and raised up over the right by being bent. Then it is extended, and the left foot is placed ahead of the right. Thus Themistius says, commenting on the third book of *The Soul*, "Animals advance by alterately pulling back and extending the left and the right."[41] As the Philosopher says in *The Forward Motion*, "When one foot is standing still, the weight is on it. But when they move forward, the leading foot must be without weight. As soon as the step has been made, this foot must again receive the weight, and the animal moves forward."[42]

Animals advance in this way, that is, by beginning motion with the right. As he says in the begining of *The Forward Motion*, "The evidence is that they carry weights on the left. For in that way it comes about that what is carried is moved more easily,"[43] when the side that should move first is unburdened. For this reason also the weight should not be placed on the moving principle, namely, the midpoint, just as it should not be placed on the side that is first moved, but only on the side that is to be moved later. For if it is placed on the mover or on the principle of motion, either the animal will not move at all or only with difficulty. Thus it is clear how, because the parts of the organ are moved in quantitative motion, they must be changed in the different motions and in shape. But because the connected parts are not changed in quantity, they need not be changed in shape in themselves, however they might be moved. Indeed, in themselves, they are related to the midpoint in a straight line – for of themselves they are straight – and thus in accord

41. Themistius, *The Soul* (*De anima*) VII, tr. Moerbeke, ed. G. Verbeke, p. 269, ll. 32-34.
42. Aristotle, *The Forward Motion of Animals* (*De animalium incessu*) c. 12, 711a21-24, 27.
43. Aristotle, *The Forward Motion of Animals* (*De animalium incessu*) c. 4, 705b29-31.

with a triangular figure. But if they are related to a curved figure, this happens to them because they are only joined to that point by the curved lines of the organ.

What has been said takes place not only in the forward motion of walking things that have clearly distinct right and left members, but also in reptiles and other animals that do not have such clearly distinct parts. As the Philosopher says in the same place, speaking of right and left, "This distinction is more clearly defined in some than in others. Those that use organic parts, such as feet or wings or anything else of the sort, accomplish the change we mentioned with more differentiation in this regard. But those which do not use such parts advance, producing distinctions in the body itself. In that way those things advance that do not have feet, such as, snakes and the so-called earthworm."[44] For what we have said applies to these things as well, though it is not equally obvious.

Hence, in a jump, where the right and the left are moved as if at the same time, the right is naturally moved first, though not noticeably so. It also happens this way in things that fly with wings or that swim with fins; they first move the right wing or fin, although not noticeably so. For this reason their motion is more like a jump than like walking.

What we have already said takes place in things having more feet, for instance, four, just as it takes place in something that has only two feet. "Because," as the Philosopher says, "their walking is mainly forward, their back members are moved forward along the diagonal; after the right front foot they move the left rear foot, then the left front foot and after that the right rear foot."[45]

Things that have more than four feet produce motion in the same way; just as in something with four feet, the back feet must be moved toward the front in second place and along the diagonal. This is clear in things that move slowly. As it happens in those things with four or more feet, so it happens in fish that have four or more fins. It would happen the same way in birds if they had four or more wings. The motions in all of them, whether of feet or wings or fins, would be toward the side, as it is in things that have two or more of them.

Something special happens in the movement of crawling things, as is obvious in the earthworm. Once it has been contracted upon the earth in its whole body toward the middle, where there is the analogue of the heart, the front parts are first extended and stretched forth by a pushing that begins in the middle, though it is first apparent in the front part. Then, in second place, the front parts are pressed down and contracted; once again this begins from the point we mentioned and is first apparent in the same place. At the same time the rear parts are extended by a pull-

44. Aristotle, *The Forward Motion of Animals* (*De animalium incessu*) c. 4, 705b21-28.
45. Aristotle, *The Forward Motion of Animals* (*De animalium incessu*) c. 14, 712a24-28.

ing from the same point that appears in the same place. After they have been extended, they are, then, in the third place, contracted to the same point, and it is first apparent there, and the whole animal is contracted as before. Then it begins to move forward again as before.

In this way, it is clear how in this forward movement of animals the mover and the moved are not distinct in their entirety and do not touch each other from the outside, as happens in the movers and the moved in the previous manner. Rather, the soul is the same in the whole body, and it is utterly simple in the case of the intellective soul in humans, although it is extended throughout the parts of the body in brute animals. Though it is indivisible, it organically moves the whole animal at a single point in the way we said. Thus it moves itself in a sense insofar as it is in the parts that it moves, and it is also moving insofar as it is in point mentioned. But it moves itself accidentally, as we have said.

The movers and the things moved at the fourth level, namely, "heavy and light things," are even less distinct. Since they are simple inanimate bodies, they are not moved by appetite, but through natural impulse, and they have no distinct organs, as is the case in animals. In no sense can they "move by themselves," as the Philosopher determined in the eighth book of the *Physics*, that is, so that they are their own principle of begining motion without an extrinsic cause, so that the thing is a mover by a principle that, while being in one part, moves another part in turn. "For this is proper to living things, and" otherwise "they could stop by themselves," in the middle of the motion, as he says in the same place.[46] With reference to this, he also says there that "those things which do not move by themselves, such as heavy and light things," are moved "either by what generates or produces the heavy or the light thing or by what removes impeding or preventing factors. From one of these they must derive the beginning of their movement."[47] When they have begun to move, they continue to move, carrying on the motion by themselves, and, as the Commentator says on the eighth book of the *Physics*, "In this respect they move by themselves accidentally. The heavy thing has a receptive principle so that it is moved from the outside, not a principle of action, save accidentally."[48] For they move by themselves in carrying out the motion, but they cannot begin to do it save through something else, either through a generating cause that gives the form by which they carry out the movement, or through what removes an impediment so that they are put in a state in which they can carry out the movement through a form they already possess. In that way they move by themselves accidentally, not because, in carrying out the movement, they are moved by

46. Aristotle, *Physics* VIII, c.4, 255a2-7.
47. Aristotle, *Physics* VIII, c. 4, 255b35-256a2.
48. Averroes, *The Commentary on the Physics* (*Phys. VIII Comm.* 32), ed. Junt. IV, 372D.

something other than themselves, nor because, when the impediment ceases, they begin to move by something other than themselves. Indeed, with regard to carrying out the motion, they are by themselves the cause of the motion, and the generator and the remover of an impediment are causes accidentally. Rather, they move by themselves accidentally, because they begin to move themselves after initially not being in motion only because they were not in a state in which they could be moved. Either they lacked a form, or there was an impediment for the form already possessed. In this respect they are the cause of their motion as accidental causes, while the generator and the remover of an impediment are the essential causes so that, if they were in their proper places, they would in no way be moved from them except by violence.

In this respect heavy and light things and animals move by themselves in different ways. Animals, which move by themselves without anything else moving them except according to the character of the desirable object, begin motion after first not being in motion and stop moving of themselves – something that heavy and light things cannot do. Hence, Averroes says in the same place, "The generating cause is that which gives to the simple generated body its form and all the accidents belonging to the form, one of which is motion in place."[49]

But understand that the generating cause does not by itself immediately bestow motion, as it bestows form. Rather, by giving form, it gives motion, because the form, when not prevented, is a sufficient cause of motion in these things. Hence, when something has the complete form outside its natural place in the absence of the generating cause, if the impediment ceases, it immediately moves to its place by itself without anything else moving it. Thus Algazel says in the first book of *Metaphysics*: "Causes are divided into essential causes and accidental causes. An acccidental cause is called a cause in an improper sense, because the effect does not come from it, but from something else which only becomes the sufficient cause of the occurrence of the effect along with it. Thus one who removes a supporting column is said to bring down the roof. That is not true, because the cause of the collapse of the roof is its heaviness. It is held back in the meanwhile by the support of the column. The removal, then, of the column renders the roof apt to fall, and it collapses, which was the proper action of it, that is, of something heavy. Or, scammony is said to cool,[50] because it removes cholera, which prevents nature from cooling. Nature then is what cools, but only when what prevents it has been removed; thus, scammony will be the cause of the

---

49. Averroes, *The Commentary on the Physics* (*Phys. VIII Comm.* 32), ed. Junt. IV, 370G.
50. Scammony is an Eastern Mediterrenean plant or a preparation made from its roots which were formerly used as a cathartic.

removal of cholera, not the cause of cooling which naturally follows after the removal of cholera."[51]

Thus it is clear that nature is the essential cause of cooling, while the removal of cholera is only its accidental cause, and that scammony is the essential cause of the removal of cholera and the accidental cause of the cooling. In the same way, in the case of something heavy the essential cause producing its descent is the heavy thing with its form, and the removal of what prevents it or that which removes what prevents it is the accidental cause of its descent. But that which removes what prevents it is the essential cause of the removal of what prevents it. On this account, Averroes adds after what we have already cited, "The potency for motion is found in a simple body as an accident. If one finds that at the time of its generation the body is not moved to its natural place because of some impediment, when it is generated outside its natural place or when it leaves its natural place because of something pulling it, it does not need an essential extrinsic mover in order to go into act, since it is in accidental potency. Since this is so, that which is moved essentially is that in which there is true potency for motion. Then that which is truly moved by this motion is the matter out of which the simple body is generated. For example, air, which is potentially fire and higher up, is that which is moved truly and essentially to a higher place, when it becomes fire."[52] This comes from the generating cause, as was said. When something has been generated and is outside its place, it is moved to it by itself without anything else, as the Philosopher says in the eighth book of the *Physics*, "It happens, if it is impeded, that it is not on top, but if the impediment is removed, it acts and always rises above."[53]

That which moves essentially is, for example, the accidental form of the heavy thing, but that which is moved essentially is the body of the heavy thing composed of matter and substantial form. These are not distinguished either in place or location, as the mover and the moved are distinguished by the parts in an animal in the case of local motion, as we have said, and also in the case of the motion of growth. They are only distinguished as the form and the subject actuated by the form, in accord with what the Philosopher says, when he assigns the difference between that which is moved from itself, which occurs in growth, and the motion of heavy and light things. He points out that in one respect heavy and light things have within themselves the principle of their motions to a greater degree than things which grow. He says in the fourth book of *Heaven and Earth*, "Heavy and light things seem to have in themselves the principle"

51. Algazel, *Metaphysics* I, 1, ed. Muckle, p. 39, ll. 22-35.
52. Averroes, *The Commentary on the Physics* (*Phys. VIII Comm.* 32), ed. Junt. IV, 370GH.
53. Aristotle, *Physics* VIII, c. 4, 255b19-21.

of these things "to a greater degree because they are very close to their substance."[54] When the Commentator explains this, he says on this passage, "In the case of something heavy, the whole moves the whole, but in the case of what grows, one part increases another distinct from it in place."[55] Elsewhere he says of that same fourth book, "The reason why the heavy thing is moved, when the impediment is removed, is that its motion follows upon the form of the heavy thing as a proper accident, and thus it does not need an extrinsic mover."[56] With regard to this he also says on the third book of the same work, "Just as the other accidents existing in a thing that has been generated exist there only by the mediation of the form of what was generated, so it is the case with motion."[57] A little later he says, "For this reason this motion is not primarily and essentially from an external mover, but from the form of what is moved. Thus something heavy is moved by itself when the impediment ceases, and in this way it is like that which is moved essentially. For a stone moves itself insofar as it is heavy in act, and it is moved insofar as it is potentially below. But the reason is that it is composed of matter and form. For its form moves insofar as it is form, and it is moved insofar as it is in matter."[58] Otherwise, it is not in potency to being below.

One should, nonetheless, notice here that, with regard to the heavy thing, motion and change come about successively and suddenly. For from the side of that which moves and changes, the heavy thing moves itself through itself, that is, through the form of the heavy thing, suddenly moving itself downward, unless there is a resisting medium, and if there is a resisting medium, it moves itself downward successively. It does this more rapidly or more slowly in accord with the quality of the medium and of the heavy thing. But from the side of that which is moved and changed, the heavy thing is changed of itself to a lower position, not through violence or the resistance of what is changed, but through obedience, and it is changed without any successive motion. They move downward by themselves, but not without a medium that resists, and in that way something heavy moves itself accidentally. Thus the Commentator says on the third book of *The Heaven and the World*, "Since that is so, it is necessary that it move itself, because it essentially moves something other than itself. For example, a man only moves himself accidentally on a ship because he moves the ship essentially, and thus a stone essentially moves only the air in which it is and thus it follows the motion of the air, as is the

---

54. Aristotle, *The Heaven* (*De caelo*) IV, c. 3, 310b31-33.
55. Averroes, *The Commentary on the Heaven* (*De caelo IV Comm*. 24), ed. Junt. V, 252b.
56. Macken suggests as a possible source: Averroes, *The Commentary on the Heaven* (*De caelo IV Comm*. 2), ed. Junt. V, 234E-235A.
57. Averroes, *The Commentary on the Heaven* (*De caelo III Comm*. 28), ed. Junt. V, 198IK.
58. Averroes, *The Commentary on the Heaven* (*De caelo III Comm*. 28), ed. Junt. V, 198KL.

case with the man and the ship."[59] At the end of the chapter he says, "It is obvious that air is necessary for the motion of the stone, and that is what we promised and explained in the *Physics*, but this place is more convenient."[60]

Hence, after heavy and light things are generated in their complete form without any impediment, they in a sense move themselves by themselves, in passing to their natural places, to a greater extent than animals are moved by themselves to the places where the objects of their desires are found. For heavy and light things do not require something at rest either within themselves or external to themselves, as the Philosopher says in *The Causes of the Motion of Animals*, "But with regard to inanimate things that are moved, someone might wonder whether one must say that they have within them something at rest and something moving and also something extrinsic to them that is at rest. But this is impossible; think of fire and earth or some other inanimate thing."[61] Besides, if heavy and light things were not moved by themselves, because they have the beginning of motion from the generating cause or from what removes an impediment, for even better reasons animals would not be moved by themselves, since they do not have the form of the appetite save from the object of desire. From that they have the beginning of motion, as has been said, so that irrational animals do not freely move themselves for this reason.

Those things which are movers and moved at the third level are found in the intellect when it understands. When the intellect has been moved by the intelligible object in the understanding of simple intelligence, by its natural power of conversion to itself and to its act and to its object, it opposes itself to itself as an intellective power naturally able to be moved by itself, as by the intellective memory informed with simple knowledge about the object. The knowledge and the object constitute one object that moves the same intellect as turned back on itself, by informing it with explanatory knowledge, as we have elsewhere explained in more detail.[62] Thus the same intellect is distinguished into mover and moved. For it is moving insofar as it is informed by the simple knowledge, and it is moved insofar as it is the bare intellect and in potency to explanatory knowledge. Thus, though it could not bring itself into first act, because it was only in pure potency with respect to the act of understanding, it can, nonetheless, bring itself into second act, because it had some actuality.

---

59. Averroes, *The Commentary on the Heaven* (*De caelo III Comm.* 28), ed. Junt. V, 199A.

60. Averroes, *The Commentary on the Heaven* (*De caelo III Comm.* 28), ed. Junt. V, 199B.

61. Aristotle, *The Motion of Animals* (*De motu animalium*) c. 4, 700a11-15.

62. Cf. Henry of Ghent, *The Ordinary Questions* (*Quaestiones Ordinariae* [*Summa*]), a. 58, q. 2.

Matter prevents this from occurring in material things, as we have explained in *The Ordinary Questions*.[63] At this level there is less of a difference between the mover and the moved than in the preceding levels.

There follows the mover and the moved at the second level, namely, in the will in moving itself to the act of willing. The present question is concerned with this.

In accord with Anselm's position and with what seems to be that of the Philosopher in *Good Fortune*,[64] some say that the first motion of the will cannot be from itself, but must be from God.

We have sufficiently explained in other *Quodlibetal Questions* that it is not so.[65] I grant that God moves all things according to his general administration and moves them in different ways according to the diversity of the things moved, for instance, heavy things downward, light things upward. Thus he moves the will according to its condition. Nonetheless, the motion of all things ought not on this ground to be ascribed to him except as to the universal cause. Apart from this, the different particular causes of the diverse motions must be examined, and their effects should not be attributed to God, although their powers come from him. In the same way the motion of heavy things should not be attributed to the cause that gives a heavy thing the form which it moves. With respect to this point, Augustine says that God administers things so that he allows them to produce their proper motions.[66]

Others say that the will is moved by the known good as a passive power by its proper object, just as the intellect is moved by the knowable truth, but naturally, not violently.

This cannot be the case. For, if it were, then, just as the intellect cannot, when the intelligible is present, fail to be moved by it to the act of knowing, so the will could not, when the known good is present, fail to be moved to the act of willing. Thus free choice would perish, and as a result, all meaning for meriting well or ill, for persuasion, deliberation, counsel, and the remaining things requisite for the virtues.

Thus some say that the form of the intellect is by itself the principle of human actions and that the will is an appetite that is merely an inclination following the form of the intellect, but because the known form is universal and indeterminately related to many, the inclination of

---

63. Cf. Henry of Ghent, *The Ordinary Questions* (*Quaestiones ordinariae* [*Summa*]), a. 36, q. 1-3.

64. Anselm, *Why God Became Man* (*Cur Deus homo*) I, c. 11, ed. F. Schmitt, II, pp. 68-69; cited by Thomas Aquinas, *The Disputed Questions on Evil* (*Quaestiones disputatae de malo*), q. 7, a. 1, arg. 8; Aristotle, *Eudaemian Ethics* VII, c. 14, 1248a24-27; also cited in Thomas Aquinas, *The Disputed Questions on Evil* (*Quaestiones disputatae de malo*), q. 6.

65. Cf. Henry of Ghent, *Quodlibet VI*, q. 10 (*Henrici de Gandavo Quodlibet VI*), ed. G. Wilson, pp. 88-126.

66. Cf. *The City of God* (*De ciuitate Dei*) VII, 30.

the will is that way too. Thus it is not necessary that it be determinately inclined to one thing, just as the will of the craftsman is not determinately inclined to a particular house.[67]

Suppose that the form of the intellect were to move the will itself by inclining it, precisely so that it would be only a certain inclination to the known good, just as the inclination downward follows upon the form of the heavy thing, so that its desire for being below is nothing but this inclination. In that case, just as the apprehended universal good moves it indeterminately, so that as a result it does not of necessity seek some particular good under it, so the apprehended particular good would necessarily move it and incline it so that it determinately and necessarily seeks it. In this way free choice would be destroyed.

Hence, others say that the act of the will, which is to will, can be "considered in two ways: in one way with regard to the determination of the act, in the other way with regard to the exercise of it."[68] It is the same way in a craftsman. If he has only the form of one house in his mind, he cannot will to make another house of another form. But in the second way the will is determined by the end, just as the craftsman who has only one form of a house can still be indifferent with regard to making or not making that house. In the first way they say that the will is moved by the intellect, because the good as known is the form specifying the act of willing and it determines that, if the will wills anything, it is necessary that it will that good. But this is not the case with the second way, because in that way its object is the good as good and, hence, the end which it can indifferently either pursue or not will. Thus it is proper to the will to move itself and all other potencies to their actions or to draw them back from their actions.

But if the intellect were the moving principle of the will with regard to any specification of the act so that the known good is said to move the will, then I ask about the specification by which the will is said to be moved. For, either the good is merely shown or offered to the will by the intellect, which receives its impression from the intelligible object, just as a proper passive potency receives its impression from its proper active cause by natural necessity. Then there is no freedom of the intellect not to receive it, except in the sense that matter can receive or not receive a form insofar as there can or cannot be an agent imprinting it. But this is not due to any freedom. Or there is some inclination produced in the will.

67. Cf. Thomas Aquinas, *The Disputed Questions on Evil* (*Quaestiones disputatae de malo*), q. 6.

68. Thomas Aquinas, *The Summa of Theology* (*Summa theologiae*) I-II, q. 9, a. 1 ad 3um; *The Disputed Questions on Evil* (*Quaestiones disputatae de malo*), q. 6.

If in the first way, the will is moved neither by the known good nor by the intellect, because nothing is moved by something else unless some impression is produced in it by the other. Thus, if the will is moved, it is moved by itself, and this is the case whether it is moved in determining for itself its act and its object, or by doing or carrying out its act. Thus, in both ways there remains full freedom of the will with respect to its act. Nor does the intellect do anything to bring the will into its act, except to show or offer the object, and it does this only as an accidental cause and necessary condition. On this account, if there is present any determination, it is the determination by which the intellect is passively determined by the intelligible object that acts upon and determines it. By this passive determination in the intellect, an object is presented to the will, and by it the will is in no way itself passively determined by the active intellect to its act of willing.

If in the second way, namely, there is some inclination produced in the will, then, either that inclination is not a volition, but some impression inclining it to will, like a weight, as a habit existing in it inclines it, or it is a volition or act of willing.

If it happens in the first way, then, despite that impression inclining it, the will remains in its full freedom of acting and not acting in accord with that impression, just as if it did not have it, although it cannot so easily will its contrary. Thus, if it is moved to will something, it is moved by itself, and this is the case both with regard to the determination of the act and of the object and with regard to the exercise of the act, as we said before.

But if it happens in the second way, that inclination is a volition so that such an inclination is nothing but a certain willing, as Augustine says in commenting on the verse of the Psalm: "Incline my heart toward your testimonies."[69] "What does it mean to have the heart inclined toward something but to will it?"[70] But when the will wills something, it carries it out unless it is impeded, and if there is not some external action to be carried out, the exercise of the act is nothing other than willing. Thus, it is not possible to claim that the intellect moves the will in the way mentioned with respect to the determination of the act and not with respect to the exercise of the act. Indeed, if it is necessitated in this way with regard to the determination, it is likewise necessitated with regard to the exercise, because it cannot not will to carry it out, "For the appetition is the activity," as the Philosopher say in *On the Motion of Animals*.[71] This will now be explained according to him.

---

69. Ps 118:36
70. Augustine, *The Commentary on Psalm 118 (Enarratio in Psalmum CXVIII)* s. xi, c. 6: *CC* XL, 1698.
71. Aristotle, *On the Motion of Animals (De motu animalium)* c. 7, 701a31-32.

Thus, if the will were moved by the object of the intellect however slightly, there could be no act of rejection concerning it. Rather, it would be necessary to carry out the act or to pursue the object to attain it. For what is once acted upon by something is always passive with respect to it and never active, though it could be active with respect to something else. In the same way, the intellect is passive with respect to the object of simple understanding, but once it has been acted upon, it can act upon itself to generate in itself explanatory knowledge and to constitute new complex intelligibles regarding the intelligibles first known. So too, insofar as the intellect is in act with regard to the knowledge of principles, it acts so that it becomes in act with respect to knowledge of conclusions. Likewise, if the will were actuated to will something by the intellect, it could well enough move itself to will something else ordered to that end, as one willing health by the motion of the intellect could move himself to will some medicine. Yet the will could not, on the basis of this position, cease from willing that which the intellect moved it to will, after it was determined for it by reason. For, insofar as it is determined, it is either the end or includes the meaning of the end, insofar, that is, as without it the end could not be attained. And thus it necessarily moves, as an end and as good with regard to all the particulars and with regard to every consideration of good. Such a mover necessitates the will by overcoming it with respect to the possibility of the contrary.

Hence, if the intellect proposes some things to the will as goods of the sort that it can propose to itself without any determination, the will is not moved necessarily. No apprehended good necessarily moves the will except the end which is good according to every meaning and consideration or that which includes the meaning of the end insofar as it was determined by reason. Hence, the Philosopher says in *The Motions of Animals*: "The object of desire and of the intellect is what first moves – not every such object, but only what is the end of action. For this reason, some goods are things that move, but not every good is. A good moves insofar as something else is the end and insofar as it is the end of those things which are for the sake of something else,"[72] and it moves of necessity.

But insofar as something else is the good, it does not move to that good, as some say, without preceding deliberation.[73] Thus, when someone wills health, he begins to deliberate about the means to health in such a way that he does not will any of them until deliberation has been completed. He can freely seek or not seek this advice from reason and either wait for or not wait for what he has sought; he can either follow or not follow the advice that he waited for and that was given by reason,

---

72. Aristotle, *The Motion of Animals* (*De motu animalium*) c. 6, 700b23-28.
73. Cf. Thomas Aquinas, *The Disputed Questions on Evil* (*Quaestiones disputatae de malo*), q. 6.

because deliberation is a non-demonstrative investigation, and they say
that the freedom of the will consists in it.

If the good were determined by reason as a demonstrative conclusion that includes the meaning of the end, some say that the will cannot
not will that. On the other hand, if it were demonstrated by reason, not as
a demonstrative conclusion, but as a persuasive conclusion, they say that it
is able not to will it. But if it does will or choose that good, another
motion of the will, not from itself but from something else, must precede
that motion of the will by which he wills it, along with the preceding deliberation. And it must also precede the motion of the will by which it wills
to deliberate about willing it. This other motion will be from some object
of desire known and determined as having the character of an end or, if it
is not from that object of desire, since this object of desire is what is first
willed, though not without preceding deliberation, it is moved by something else, such as a heavenly body or God or fate or good fortune or
something of the sort.

The first of these alternatives, namely, "if it were a good
determined by reason without qualification and thus as a demonstrative
conclusion and as including the character of the end" so that "the will
could not not will it,"[74] but is moved to willing this by the known good, as
a naturally passive potency is moved by its proper cause, seems to be the
position of the Philosopher in the sixth book of the *Ethics*.[75] It is based on
the idea that the appetite is a mover moved by the known object of desire,
as he determines in the third book of *The Soul* and in the book, *The
Causes of the Motion of Animals*.[76] There he claims that the will is moved
of necessity to consent or to will and to pursue what it wills, if it is not
externally impeded, just as a passive potency is moved by the known
object of desire as by its proper cause. In the same way, the intellect is
moved to assent to or to affirm or to insist upon what is known by the
known truth as by its proper cause. And so, just as the force of a demonstration moves the intellect in speculative matters so that it cannot dissent
from the conclusion, so it moves the will in matters of action. He seems to
me to state this opinion more clearly in *The Motion of Animals* than elsewhere. There he says, "As the intellect at times acts and at other times
does not act, and at times moves and at other times does not move, the
same thing seems to happen also in those that know and reason about
immobile beings. After all, there the end is contemplation. For, when one
knows the two propositions, he has understood and formed the conclusion. Here, however, the conclusion from two propositions is the action.

---

74. Cf. above, p. 54, para. 1.
75. Cf. Aristotle, *Nicomachaean Ethics* VI, c. 2, 1139a17-b4.
76. Cf. Aristotle, *The Soul* (*De anima*) III, c. 10, 433a9-b30 and *The Motion of Animals* (*De
   motu animalium*) c. 7, 701a7-b31.

For example, when one has understood that every man should walk, the man himself immediately walks. But if no one should now walk, he is immediately at rest. And he does both of these if nothing prevents him."[77] There he adds after some intervening material, "It is clear that the action is the conclusion. Desire says that I should drink; the senses, or the imagination, or the intellect says that this is something to drink; one drinks at once."[78] In this case he makes no distinction between the motion of the rational appetite by the intellect and that of the sensitive appetite by sense or imagination; this is clear from what he immediately adds, "In this way animals begin to move and to act; the ultimate cause of their motion is appetite. This is produced either by the senses or by the imagination and intelligence of those who desire to act. At times this is on account of desire or anger; at times on account of will. At times they produce something; at times they do something."[79]

Although they hold the first of the alternatives we mentioned above, namely, that, "if the good is determined by reason as a demonstrative conclusion," they say that the will is moved by the intellect without violence or coercion. After all, it is not moved as something naturally determined to the opposing contrary, as something heavy is moved upwards. Rather, as something that is indifferent to many things, it is moved by something else that determines it to one of them. Nonetheless, they completely remove freedom of choice in willing the object of the will, because it requires freedom from all necessity. Accordingly, in the case of God, in the breathing forth of the Holy Spirit, we say that he is breathed forth by the Father and the Son with free will; nonetheless, on account of their necessary immutability, we do not say that he is breathed forth by the free choice of the will, as we say that creatures are created by God. On this subject enough has been said elsewhere in the *Quodlibetal Questions*, namely, that the will cannot be determined in this way by anything the intellect has determinated apart from the highest good when it is immediately seen.[80] Hence, in this life the will cannot be determined except in the universal insofar as a man cannot not will to be happy. This does not pertain to the present question which concerns only whether the will, in so willing, is moved by something else. After all, at the presence of the object of desire in the intellect, without any change which the will receives from the known object of desire, the will, in freely moving itself to the act of willing, is carried toward the object of the will. But if there is added immutable necessity, even without any coercion, so that it cannot

---

77. Aristotle, *The Motion of Animals* (*De motu animalium*) c. 7, 701a7-a16.
78. Aristotle, *The Motion of Animals* (*De motu animalium*) c. 7, 701a22-23 and 32-33.
79. Aristotle, *The Motion of Animals* (*De motu animalium*) c. 7, 701a33-b1.
80. Cf. Henry of Ghent, *Quodlibet* I, q. 16 (*Henrici de Gandavo Quodlibet I*), ed. Macken, p. 100, ll. 33-37, 44-49; p. 113, ll. 17-23.

not will what it so wills, freedom of choice will be completely done away with, as we have said.

That, in so willing, the will is not moved by the object known is clear from the fact that the rational appetite, called the will, would, in that case, be moved by the object of desire known by the intellect with the same necessity by which the sensible appetite is moved by the object of desire known through sensation and imagination. The Philosopher has put this point well, as will now be said.

This position is false, because, on this view, the will would not be rational, properly speaking. Not only would it not have freedom of choice, but it would be acted upon rather than acting, as Damascene says in chapter twenty of the second book of the *Sentences*, "In irrational beings there is produced an appetite for something, and immediately there is an impulse toward action, and the animals are driven by appetite. For this reason the appetite of irrational animals is not called will. For in humans, rational beings, the rational appetite leads rather than is led."[81] And in chapter twenty-nine he says, "Either one will not be rational, or, as a rational being, one will be master of one's actions."[82] Further on, he adds concerning irrational beings, "They are acted upon by nature rather than act, and they do not resist natural appetite. Rather, as soon as they desire, they are impelled to act. A human being, a rational being, acts upon his nature rather than is acted upon."[83]

The second of the aforementioned alternatives was that "if the good has not been determined by reason as a demonstrative conclusion, but rather as a conclusion persuading one to will it or to will to deliberate about it, another motion of the will from something other than it must precede it, even though one does not assert that the other motion comes from a good known in its character as end."[84] They try to explain this by the following argument: "If someone wills one of the means to the end, after previously not willing it, and does not will it of necessity, as having been moved by the intellect, he does not will it without preceding deliberation.[85] He does not undertake or seek deliberation necessarily, but freely; in fact, he does so by the free choice of the will willing to deliberate on this matter. Since, then, the will is only moved by deliberation in those things which are means to the end, the will, which now wills to deliberate and previously did not will, is necessarily moved by something so that it wills to deliberate before it wills that which is to be willed as the result of deliberation. Hence, it is moved to will to deliberate either by

81. John Damascene, *The Orthodox Faith* (*De fide orthodoxa*) c. 36: *PG* 94, 946C.
82. John Damascene, *The Orthodox Faith* (*De fide orthodoxa*) c. 41: *PG* 94, 962A.
83. John Damascene, *The Orthodox Faith* (*De fide orthodoxa*) c. 41: *PG* 94, 961A.
84. Cf. above, p. 54, para. 1.
85. Cf. Thomas Aquinas, *The Disputed Questions on Evil* (*Quaestiones disputatae de malo*), q. 6.

itself or by another. It is not moved by itself, because, since deliberation concerns those things which are means to the end, deliberation necessarily precedes this motion of the will, and by the same reasoning something else precedes that, and so on to infinity. Since this is impossible, it is necessary to hold that by the first motion the will is moved to will the means to the end and whatever is to be willed by counsel from another than oneself."[86]

The Philosopher really stated that this movement was either from God in those things which do not fall under free choice and which are not in the power of our knowledge or foresight, as we explained in the question on good fortune,[87] or from the celestial bodies in these things which can fall under our natural apprehension from the senses, but through the mediation of the intellect. He states that the intellective appetite is determined and moved by the impression of the celestial bodies, just as the sensitive appetite is. He said this of these things in which the will is determined by the intellect, because he did not maintain a determination in the will from the intellect with regard to all things known. In accord with this, he says toward the end of *The Motion of Animals*, "Sensations are, from the first, certain modifications. But the imagination and the intellective power have the power of real things. For in some way the known form of hot or cold, of the pleasant or unpleasant is like each of these things. For this reason those who merely think of them tremble and are afraid. All these things are affections and modifications in the body. But when these changes occur in the body, some are great, some are less."[88] Further on he says, "The principle of movement, as has been said, in things to be done is the object of pursuit or of flight. Warmth and coldness follow of necessity upon the pondering and imagination of them. For what is painful is something to avoid, and what is pleasant is something to pursue. But painful and pleasant things are almost all accompanied by a coldness or warmth."[89] Further on he says, "Since these things occur in this way, being acted upon and acting are naturally such that the one is active and the other passive. If neither of them falls short of its definition, the one immediately acts and the other is acted upon. Accordingly, one knows that he should walk and immediately, so to speak, walks, if nothing else prevents him. For, the appropriate affections command the organic parts, and the appetite commands the affections, and the imagination commands the appetite. But this latter occurs either

---

86. This long explanation, which Henry presents as a citation, may reflect the position of one of those who made an oral objection at the quodlibetal disputation.
87. Cf. Henry of Ghent, *Quodlibet* VI, q. 10 (*Quaestiones quodlibetales*), ed. 1518, f. 226r-231v; ed. 1613, I, p. 344va-35rb.
88. Aristotle, *The Motion of Animals* (*De motu animalium*) c.7, 701b17-24.
89. Aristotle, *The Motion of Animals* (*De motu animalium*) c. 8, 701b33-702a1.

through the intellect or through the senses. The process is simultaneous and quick, because the passive and the active are by nature correlative."[90] On this view, the saying, "The will in human beings is such as the father of men (that is, heaven) brings about on that day," may be understood as holding true of the motion coming from the heaven mediately. The Philosopher introduces this point against those who held that the intellect is moved by heaven immediately and that the intellect is nothing but the imagination.[91] Nor is this surprising; after all, sensible things are the effects of the heaven, and the senses are formed in accord with them, and the intellect in accord with the senses, and the will in accord with the intellect, and in this way the natural active and passive principles are formed. Even if the soul is said to be an incorporeal power, still, in saying that its will is naturally moved by the intellect, one says that it is moved by the heaven, though not immediately and directly. Hence, in the third book of *The Soul* the Philosopher does not distinguish the moving principle which is the will from the intellect, but includes it under the intellect. He also links the imaginative power with the intellect in man as the same principle, but he distinguishes the imaginative power from the sensible appetite. In this way, by distinguishing the imagination from the intellect, he holds three movers and yet, by counting them as one, he has only two. Thus he claims that a man is moved only if the practical intellect and the sensible appetite agree, and he claims that, when they do agree, he is necessarily moved.[92] At times they move separately, when one appetite conquers another. Since the intellect does not move save by reason of its appetite, he determines that the motion arises from these two, not "in their diversity," but in the common character in which they agree. For, otherwise, as the Commentator says, "motion would come from them only accidentally."[93]

As had been said, if the will were naturally moved by something else, it would be determined to its act without any freedom, and it could not pull back from it. Thus it would not be "the master of its own acts," nor would the appetite which is the will "have the power to restrain the appetite" in those matters which fall short of the vision of the last end. Damascene states the opposite of this in the twenty-ninth chapter that we already mentioned.[94] One must say, then, without qualification that the will is moved to its act of willing by nothing else, but is moved by itself alone.

----

90. Aristotle, *The Motion of Animals* (*De motu animalium*) c. 8, 702a10-21.
91. Cf. Aristotle, *The Motion of Animals* (*De motu animalium*) cc. 3-4, 699a12-b13.
92. Cf. Aristotle, *The Soul* (*De anima*) III, c. 10, 433a3-b30.
93. Averroes, *The Commentary on the Soul* (*Comm. in de anima*) 50, ed. Crawford, p. 519.
94. John Damascene, *The Orthodox Faith* (*De fide orthodoxa*) c. 41, *PG* 962A.

The point assumed in the argumentation already stated, namely, that "the will is not moved in those things that are means to the end except by deliberation,"[95] must be declared false. In fact, without any deliberation determined by reason to one alternative, it can move by itself toward any good proposed, short of the last end when it is clearly seen. It can do so without movement from anything else, just as it can turn aside from it, and in this way the rest of the process does not hold.

We have, then, to consider how we should state that the will is moved by itself to the act of willing. After all, the will is a higher power than the intellect and, consequently, than all those things that move in the ways we have already treated, and on account of its freedom there is nothing higher than it save God. Hence, we should expect less of a difference between it and what is properly moved by it than in the movers and things moved in the ways that preceded, though there is a greater difference than in God. In God, there is no difference in his act of understanding between the mover, that which is moved, and the very act by which it is moved, except merely a difference in our minds, as has been said.

One should know, then, that in the will one can consider that, though the potency or power that receives into itself the act of willing and freedom without qualification and freedom of choice belong to the same potency, they do not differ merely in the way we think of them, but as powers of the potency. This arises from the nature of the will, and not merely from the way reason looks at it.

Let us understand the will, insofar as it is a potency, to be passive and in potency to the act of willing, just as the intellect is in potency to the act of understanding. It is necessary to admit this, by whatever mover one holds that the will is moved when it begins to will after not willing. After all, its willing is an accident and activity in which lies its perfection in well-being.

On account of this freedom of the will by which it ought to be the master of its actions, it is impossible to hold that, as a nature and as receptive of the act of willing, it proceeds from potency to act by some natural active principle other than itself. This is, after all, utterly incompatible with freedom, as has been said, and the appetite of the will would be acted upon no less than the sensible appetite. Accordingly, if the Holy Spirit proceeded from the Father and the Son as from a natural active principle, as a nature and in the manner of a nature, he would not be said to proceed in the manner of the will and of freedom or of liberality. But because he proceeds from an agent insofar as the agent is free in will, even though natural immutable necessity accompanies it, he is said to proceed by an act of the will. Likewise, if under God's action the will was

---

95. Cf. above, p. 50, para. 4

directly changed so that it began to will after not willing, even though this were not contrary to its nature, but only apart from it, the act would still not be called free, nor would it be praiseworthy on this view, nor an act of virtue at all. After all, since its action should be present in it in such a way that the will is said to be praiseworthy and virtuous in accord with it, it must be directly moved to the act of willing by nothing else than by itself alone, although with the assistance and cooperation of another, for instance, the Holy Spirit, either through himself and immediately, or by some gift of his, or in both ways. Accordingly, we hold that the act of willing or loving the God of glory in the clear vision of God is elicited only by the will, with the cooperation of the Holy Spirit in himself and in his gift.

Some wish to maintain that the motion of the will follows the information of the intellect in the way explained in the activities already discussed, although they do not claim that it is moved by the intellect. They would say that, insofar as it is free, the will moves itself insofar as it is a potency that is receptive of the act of willing, but only in accord with the information of the intellect, and that this is natural for it. In accord with this view, the previously mentioned ways could be explained, except for the intellect's not being said to move the will.

Let us suppose that that was the case and that we could, contrary to the opinions mentioned, avoid the contradictory conclusion that the will was not the master of its acts and was rather acted upon than acting. After all, in making that assertion, we claim that the will is the master of its acts and we say that it acts rather than is acted upon. It still would not be possible to avoid the complete exclusion of the freedom of choice by which the will must be borne without any necessity by its act to its object. We could only do this by claiming that the will can will what it does not will and can not will what it does will to the extent that the intellect can not understand what it understands or can understand what it does not understand. Thus the will's freedom of choice would depend upon reason and should be attributed to reason rather than to will.

Thus others would still say that the will, in as much as it is free, inclines itself to the known good insofar as it is known,[96] and that this is a certain willing, albeit imperfect, as is clear from what we have already said, following Augustine.[97] Thus the will would be moved by itself with a certain immutable necessity to the known good that has the most weight, especially after the determination of reason, but not with such great adhesion to it that it cannot by itself, as a matter of free choice, reject or pursue that act.

---

96. The position of these unidentified persons may reflect the tenor of some of the objections that were orally posed by persons present at Henry's quodlibetal disputation rather than objections to be found in some written source.
97. Cf. above, p. 50, para. 5.

Though such an act is freely elicited by the will, it still takes place with a certain immutable necessity in which there can be neither moral praise nor blame. It would, then, be indifferent, neither praiseworthy, even if it were directed toward a known good which is good without qualification, nor blameworthy, though it were directed toward a pleasureful and merely apparent good. But the first motions of the senses are, according to some, judged to be blameworthy without qualification, even though only slightly blameworthy, so that they are not imputed to the point that punishment is required for them, or so that there is only required the punishment that is owed to venial sin. This seems quite contradictory. Hence, it seems that one should not hold this view. Rather, our practical intellect, insofar as it is practical, does something more with regard to the will than the speculative intellect does. Thus, when it discloses a doable good in its character as doable, something happens in the will that does not happen in it when the speculative intellect discloses something true that is not doable in its character as true. Otherwise, the Philosopher would have no more considered the will together with the practical intellect rather than with the speculative intellect as a single moving principle, nor would the practical intellect be called a moving principle rather than the speculative intellect. I claim that, though the will is in no sense moved by the practical intellect in terms of the act of willing, changing it from not willing – I do not mean: from unwilling – to willing, it is, nonetheless, moved by it in terms of a modification which is like a weight in the will, which is still free. It inclines it to will in the manner of a habit, just as in the sensitive appetites modifications in terms of hot and cold cause the appetites. But those appetites are rather acted upon than acting, and irrational animals are acted upon in accord with them. Let me give a better example. A grace from God in the will inclines it as a weight so that it wills, but it does not elicit the act of the will nor compel the will to elicit it. The act can only be elicited by the free will, though frequently it would not elicit the act without such a weight inclining it. Furthermore, in matters pertaining to merit it could in no way meritoriously elicit the act without that weight.

As a second point, this "inclination of the will" is equivocal. In one way, it is the incomplete will by which it is imperfectly moved to the thing, and about this we said above, following Augustine, that it is nothing other than willing.[98] In another way, it is the weight of the modification from the practical intellect or of some infused gift or perhaps of some impulse produced in the will.

---

98. Cf. above, p. 50, par. 5.

In the first way, something heavy is said to be inclined downward when it begins to move, even slowly.

In the second way, something heavy is said to be inclined downward when, while resting on top, it presses down on that upon which it is resting in order to be able to move downward. Hence, when the force of the weight is stronger in pressing down than that of the support in holding it up, it drives out its support and, by removing the impediment, comes down.

The Philosopher claimed that this weight in the will is the good itself that is known, as we have explained.[99] Thus the animal would always be moved according to the intellect unless there were an opposing appetite of greater weight. He stated that this opposing appetite could be of such great weight that it completely overwhelmed the motion of the will so that at times it could not resist in any way.

This is true, unless one is helped by grace, although some heretics have claimed that the will was by itself sufficient to repulse every motion arising from the sensitive appetite. It is true that, if the sensitive motion has enough weight to overcome the motion of the will, it does this, not by doing violence to it, but by enticing it by such a great weight that the will must assent to it, though freely, or seek for help from above by doing what lies in its power. If it did this, the help would not be lacking to it. But, if it does not do this, once having been enticed, it consents with the free choice of the will. Yet the intellect alone cannot impose such a violent weight upon the will; only its own wickedness can do so.

Accordingly, in seeing the small amount of weight that the intellect can by itself impose upon it, I state that it is not so great that the will, free as it is in its choice, cannot will the opposite of that to which the weight inclines it, no matter what it is, short of the clear vision of the ultimate good that is our end, and no matter what the intellect has determined should be done or willed. Only the good that is our end known by the intellect with clear vision has so much weight inclining the will – by enticing it, not by doing it violence – that it freely wills that good, though it cannot, nonetheless, by a certain immutable necessity, not will it. I say that it wills it freely, but not by free choice. Properly speaking, it is not free choice, but only freedom that has to do with the end; free choice concerns means to the end. According to the Philosopher, free choice concerns the means, only insofar as its choosing depends upon the judgment and determination of reason, as has been said.[100] In accord with this, Damascene distinguished in the twenty-sixth chapter of the second book between something that we do voluntarily when we choose and

---

99. Cf. above, pp. 57-58.
100. Cf. above, pp. 57-58.

something that we do voluntarily without choosing.[101] The will produces the first sort of willing without reasoning and, hence, quickly; it produces the second sort, according to the Philosopher, only with reason and, hence, not quickly, but only after reason has made its decision. When the practial intelligence proposes the major premise as a determinate good and the minor premise is known to the speculative intellect, then, given the major, the activity follows in the place of the conclusion, without pausing over the minor. This is done with free will, because the good which entices the will is, in the major premise, attractive, but freely so, even though without any choosing by free choice, because it does not wait for the minor premise. In the beatific vision, when the highest good is seen, the practical intellect says, "To will this is good for a human being." But by the speculative intellect each one knows that one is a human being, and thus without consideration of the minor premise, once the highest good is seen, the will quickly and immutably wills it without any reasoning and consideration of the minor premise.

When the appetite proposes in the major premise a doable good as something possible to be determined and when sensation, imagination or intellect has to determine the minor premise by knowledge, then, given the major, the activity does not, according to the Philosopher, follow without a consideration of the minor. Thus the will wills the conclusion, not merely freely, but by free choice, and it does not will it quickly and without reasoning. Hence, the Philosopher says in *The Motions of Animals*, "Practical propositions are formed in two kinds: in terms of what is good and in terms of what is possible. For example, if walking is good for a human being, the intellect does not delay over the fact that one is a human being. For this reason whatever we do without reasoning, we do quickly," as happens with the sensible appetite which is called desire. "It says, 'Something to drink is good for me.' The senses or imagination or intellect say, 'This is something to drink.' And one drinks immediately," but not without previous reasoning or consideration of the minor premise.[102]

We do not follow the Philosopher on this last point for two reasons.

First, with regard to the first kind of practical proposition, although a practical proposition is proposed as good, except in the case of a clear vision of the highest good, the will does not necessarily act immediately; in fact, it can freely reject what is proposed.

Second, with regard to the second kind of practical proposition, although a practical proposition is proposed as possible and although sensation, imagination or intellect provides the minor premise, the will does

---

101. Cf. John Damascene, *The Orthodox Faith* (*De fide orthodoxa*) c. 38: *PG* 94, 958B-C.
102. Aristotle, *The Motion of Animals* (*De motu animalium*) c. 7, 701a23-33.

not in that case act necessarily. For, although such an action is determined for the will by such a demonstrative conclusion, it can freely reject
it, even though there is imposed upon it a weight by which it is inclined to
do it and by which conscience begins to be formed for the will that it
should do that action, which, despite conscience, it can still not do. The
will cannot reject this weight directly. Once it is naturally passive with
respect to this, it is always passive with respect to this, and it is not active
either in increasing such a weight or in directly rejecting it. Nonetheless,
the will itself, which is first in itself passive because of the reception of
such a weight, need not always be passive with respect to everything and
active in no way; in fact, insofar as it is passive with regard to such a disposition, it becomes more effectively active through it for producing in
itself another disposition. For, even though the will can of itself move
itself to the known good in terms of the act of willing without any weight
inclining it, though not so efficaciously, it is also active, in the second
place, for the rejection of this weight indirectly. For, although that weight
was imposed upon it by reason, because a path of reasoning determined
that the good was to be willed, the will can by its command direct reason
or intellect to find an equally efficacious reason to the contrary. Or, if it
cannot, it could compel it to believe the contrary, at least if reason is not
determined by a truly demonstrative and evident minor premise. Yet, it is
not compelled to believe that without some sort of reason. Or it can be
said, as we have elsewhere determined: even if such a weight were not in
the will as the result of a determination of reason, the will could still by
itself move itself to whatever it wills, short of the ultimate end.[103]

Accordingly, we must reply to the question that the will alone
moves itself to the act of willing, in accord with what has been said. It
moves itself freely to the highest good seen as present and to the same
highest good when it is not seen as present except in the universal, just as
in the universal no one cannot will to be happy. But it moves itself
through free choice to anything else.

It fittingly has such an active power toward itself, because, insofar
as it is separated from matter, it is able to return to itself in acting upon
itself, just as the intellect is able to return to itself in being acted upon by
itself. The active and the passive elements are here much less distant than
in the case of the intellect, because the intellect cannot be active unless it
first is acted upon. As a result, it does not do something to itself without
that disposition, as we have explained in the *Ordinary Questions*.[104] But,
even if the will were not first passive in receiving the weight mentioned, it

---

103. Cf. Henry of Ghent, *Quodlibet* I, q. 16 (*Henrici de Gandavo Quodlibet I*), ed. Macken,
    pp. 100 and 113.
104. Cf. Henry of Ghent, *The Ordinary Questions* (*Quaestiones ordinariae* [*Summa*]) a. 48,
    q. 2.

could, nonetheless, at the mere presence of the known good in the intellect, move itself in terms of the act of willing it.

This is the case, because it does not move itself as the one who principally wills, but in the respect in which it moves. For the one to whom the potencies of intellect and will belong understands by the intellect as by something of his own, and likewise he wills by the will. Thus he understands the good by the intellect, and by the will he moves himself in the will to will it. If the intellect and will were not potencies of the same person and were not in the same substance of the soul or of the nature based on the body, but were considered as diverse things so that one of them would be what by itself principally understands and the other would be what by itself principally wills, the will would in no way move itself to the act of willing, since it is not the function of the will to know and it is not possible to will something unless it is known.

<With Regard to the Objections>

It was first argued that "the will cannot move itself, because, since it is simple, the same thing would in that case be in act and in potency with respect to the same thing."[105] One should say that it is impossible for the same thing to be in act and in potency with respect to the same thing in every way; in fact, there must always be some difference between them, but that difference is of a different sort in accord with the different definitions of the motions, the movers and the things moved. Thus there is no need for the same difference between the mover and the moved in spiritual and corporeal things, just as there is not the same kind of motion in these things and in those. Hence, as is clear from what has been said, there is in the will a sufficient difference between the mover and the moved in the sort of motion that willing is, although it is simple in reality.

Though the argument to the contrary has to be conceded on account of the conclusion, the minor premise concerning the motion of the heavy and the light by themselves and of animals in terms of place is not relevant to the question under discussion except insofar as in their case the mover and the moved are not completely different and separated from each other, just as they are not in this case. Still there is a very great diversity since in their case that by reason of which the mover moves and that by reason of which it is moved are really different. For the heavy moves itself according to form and is moved according to matter. It is not that form moves and matter is moved; rather, form is that by which the whole moves, and matter is that by which the whole is moved. Likewise, in

---

105. Cf. above, p. 32, para. 1 of q. 5.

the animal the soul is that by which it moves as form, and some part of
the body is what moves and is moved as an instrument, as parts next to
the principle that formally moves. The whole rest of the body is moved
and not moving, as it clear from what we have said. But in the present
case the mover and the moved differ only by a distinction of reason and
also by an intentional distinction. They do not differ as distinct potencies,
but as powers of one potency. Hence, the minor premise that they mentioned is of no help to prove that something that is utterly simple in reality
can move itself. In fact, it rather favors the opposite conclusion. After all,
if the heavy and the light and animals move themselves because it is possible to find in them a real diversity by which the mover and the moved can
be distinguished, then, on the contrary, since such diversity cannot be
found in the will, it follows that it can in no way move itself. And that is
true of the type of motion by which those things move themselves.
Conversely, if the will can move itself because it is a power that can bend
back upon itself on account of its simplicity so that the mover and the
moved are utterly the same in reality, it follows that the heavy and the
light and animals, since they are material, can in no sense move themselves so that the mover and the moved would be utterly the same in reality. Thus, if the will itself or that to which it belonged, for instance, a soul
or an angel, contained matter and was composed of matter and form or of
any other really distinct components, it could in no way move itself as it
now does.

Quodlibet IX, Question 6

Is commanding an act of the will or of the reason or intellect?

On the third point it was argued that to command is an act of the
reason, not of the will, because to command is merely to indicate to
another that something should be done. But that is the function of reason
with regard to the will. Therefore, and so on.

On the contrary. To command always pertains to what is highest
and free and possesses the greater dominion. In the whole kingdom of the
soul, the will alone is such. Therefore, and so on.

<The Solution>

Our answer will be the following. Since to command is an action
directed to someone in order to carry out something, one has to examine
to whom the action of commanding belongs through a comparison of
three elements to one another, namely, from the relation which the one
commanding ought to have to him to whom the command is directed, and

from the condition of the act which is commanded, and from the disposition of the one to whom the command is given. When we have examined these, it will be perfectly clear that the act of commanding ought to be attributed to the will and not to the intellect.

First, then, that to command is an act of the will, not of the intellect, is seen from the relation of the one commanding to the one to whom the command is directed. After all, with respect to the one to whom the command is directed, the one commanding ought to have the relation of a superior to an inferior at some level. An equal has no command over an equal; much less does an inferior have such power over a superior. The question concerns commanding without qualification in a human being, both with regard to what is within him, insofar as there should be some power which commands the rest, and with regard to what lies outside of him. For, if there is in a human being some power which commands the other powers which are within him, the command for those things which are outside of him should also be attributed to it. It is necessary to admit in a human being one such power that rules over the rest. The Philosopher teaches this in the first book of the *Politics*, "In all things which form a composite whole and which are made up of parts, whether continuous or discrete, there comes to light a ruling and a subject element."[1] This holds only with respect to some activity which the ruler should command the subject to perform. For this reason, Aristotle said just before, "Where the one rules and the other is subject, they have some activity."[2] In this regard, then, the present question depends very much upon the question about which power or potency is higher in man. After all, rule or command or dominion ought always to be attributed to the higher and more important. As is said in the first book of the *Politics*, "By nature the ruler is better than the ruled."[3]

As to which of the potencies in man is higher and more important, the only question concerns the intellect and the will. For this reason, those who say that the intellect is the higher say that to command belongs to the intellect and that it belongs to the will to obey and to receive the command. "For the one who commands," as they say, "directs him whom he commands to do something," and he does this "by indicating or declaring" so that to command is nothing other than "to direct." Or, it is the act "of reason" directing another to do something by some indication, not by means of counseling or persuading one to will the act. He does not do this so that the direction is expressed by a verb in the indicative mood, saying: "This is to be done by you." He does this, rather, by means of a compelling order to carry out the activity insofar as "the

1. Aristotle, *Politics* I, 5, 1254a29-31.
2. Aristotle, *Politics* I, 5, 1254a28.
3. Aristotle, *Politics* I, 5, 1254b13-14.

direction is expressed by a verb in the imperative mood, saying: 'Do this.'"
In that way, when reason perfectly commands the will to will, the will
already wills. The fact that it at times commands and the will does not will
is due to the fact that it commands imperfectly as when reason wavers
between two choices. "Thus to command is the function of reason," as
they say, "because to direct is the" proper "act of reason." As reason is
not present in brute animals, so neither is command, although "the first
mover toward the exercise of" this "act is the will," just as it is "the first
mover in all the powers of the soul for the exercise of" their "acts." Thus
command "comes about by the power of the will and with presupposition
of its act, because the second mover moves only by the power of the first
mover and the power of the prior act remains in the subsequent act. Thus
that reason moves lies in the power of the will, and to command is an act
of reason that presupposes an act of the will. By the will's power reason
moves one by its command to the exercise of the act," though not to the
determination of the act that is commanded.[4] In the same way they say
that the will moves the intellect "to the exercise of the act" by which the
will wills something, although it does not move it "to the determination of
that act," but rather just the opposite.[5] In that way, as they say, reason
could direct the act of the will, and as it also can judge that it is good to
will something, so it could command that the will will it. Thus the act of
the will could be commanded by reason. Likewise, its own proper act
could be commanded to it. For reason, insofar as it reflects upon itself, is
able by commanding to direct its own act, just as it can direct the acts of
the other powers.[6]

Supposing, however, from other questions that the will is a higher
potency than the intellect,[7] I say that, insofar as it holds the higher position, one should rather claim that it is the function of the will to command
and that it is the function of the intellect and the other potencies to obey
and receive the command. For the will can will, even contrary to the dic-
tate of reason, and can force reason to depart from its judgment and thus
to agree with it, and it can constrain all the other potencies by its power of
command. An act of the intellect must precede the will's act of commanding, since we cannot will what is unknown. By inclining the will to command in a way it has determined, the intellect determines for the will what
it is that is willed as by an indication which is not a command, but a disposition toward a command. It cannot be called a command or injunction,
because the will cannot be constrained to that motion by the intellect, and
this being constrained is, as we will soon see, necessarily required in one to

---

4. Cf. Thomas Aquinas, *The Summa of Theology* (*Summa theologiae*) I-II, q. 17, a. 1, c.
5. Cf. Thomas Aquinas, *The Summa of Theology* (*Summa theologiae*) I-II, q. 9, a. 1, ad 3um.
6. Cf. Thomas Aquinas, *The Summa of Theology* (*Summa theologiae* I-II, q. 17, a. 1, c.
7. Henry of Ghent, *Quodlibet* I, q. 14, pp. 25-29.

whom the command is given. If the will were truly constrained, there would be no other command than the indicating motion of reason, and it would be not only a motion, but compulsion to obey.

Because by its indicating motion reason is in this way a disposition toward the true command of the will, its own motion, whether it compels or not, is still called an injunction, even though it cannot be properly called a command. This injunction has part of the meaning of command, and yet true command is attributed only to the will according to the words of Damascene in *The Two Natures in the One Person of Christ*, "We call the very rational appetite an activity, because it is free in choice; it has in its power the irrational passions, and judges, governs, and restrains these: anger, desire, sensation and forward motion. For these irrational passions obey and can be persuaded by reason. Their nature is to be persuaded by reason and to be subordinate. Reason moves them as it enjoins in those things which are in accord with human nature."[8] Further on he adds, "Therefore, the intellective power and appetite are united in man so that we, as rational beings, are not acted upon against our will as irrational beings are."[9] Note that in human beings he distinguishes appetite and the intellective power and that he attributes to appetite governance and restraint – the sort of thing that is the function only of one who commands in the full sense. But he attributes to the intellective power an enjoining by persuasion in accord with reason, but not by compulsion toward the act as in the case of the will. In accord with this manner of speaking, one can explain what the Philosopher said about the speculative intellect in the third book of *The Soul*, "It does not enjoin one to fear,"[10] implying that this pertains to the practical intellect. But that is not true, because, as Themistius says on that passage, "The practical intellect is not the master of motion."[11] The master of motion is that whose function it is to order and to prescribe in the full sense, and thus the intellect does not properly say, "Do this," except by way of counsel. Nor does it have the function of directing in the full sense, but only by counseling.

In execution and commanding, it is the function of the will alone to direct, while it is the function of the intellect to be subordinate, as Damascene says, and this takes place in the way in which enjoining belongs to it, as we have said.[12] In commanding, the will is not only the first mover with regard to the exercise of the act to be carried out by the intellect so that it goes so far as to determine what should be be com-

8. John Damascene, *The Two Wills in Christ* (*De duabus in Christo voluntatibus*) c. 18, *PG* 95, 147A.
9. John Damascene, *The Two Wills in Christ* (*De duabus in Christo voluntatibus*) c. 18 bis, *PG* 95, 147B.
10. Aristotle, *The Soul* (*De anima*) III, 9, 432b31.
11. Themistius, *The Soul* (*De anima*) VII, tr. Moerbeke, ed. G. Verbeke, p. 263, ll. 4-5.
12. Cf. John Damascene, *The Orthodox Faith* (*De fide orthodoxa*) c. 36, *PG* 94, 946C.

manded; it is also the proximate mover with regard to the imposition of the act upon those powers by which the command should be carried out, although this is done in some sense by the power of the intellect, as we said. In accord with this, the act of the will can in no way be commanded by reason, just as no other act, whether one's own or another's, can be commanded by reason. Rather, whatever is commanded within the person or without is commanded by the will, and this belongs to it, in accord with what we have already said, as the result of its relation to the intellect and also to all the other powers of the soul, because it is superior to all of them. This is the first way by which commanding is seen to be an act of the will, not of the intellect, namely, from the relation of the one who commands to the one to whom the command is directed.

Secondly, the same thing is seen, in part from the relation of the one whose function it is to command to him to whom the command ought to be directed, and in part from the condition of that which is commanded. For the one commanding ought to be higher than the one to whom the command is given so that he can somehow compel the inferior to carry out what is commanded, because obedience in the lower corresponds to command in the higher. If it does not naturally correspond to it, the higher does not have command over the lower by a natural order.

At times the obedience in the lower does not correspond to command, because the commanded act is not in its power. For the intellect cannot obey the will if the will bids it to understand what is beyond its power, such as supernatural truth, or if it bids it to disagree with the conclusion of a clear demonstration, though by its command the will can keep it from thinking about it. Accordingly, the will can command nothing to the vegetative powers, because they do not have a nature that can obey.

At times obedience in the lower does not correspond perfectly to command, because it is partly in its power and partly not. In this way the sensitive appetite depends in part on the disposition of the organ, and thus it is not in its power to obey the command of reason – and on this account, it at times cannot be anticipated by reason – but in part it depends upon the strength of the soul, and thus when the will insists and reason persuades, it has to obey. For this reason, the Philosopher says in the first book of the *Politics*, "Reason has dominion over the irascible and concupiscible appetites with political rule, which is characteristic of a father regarding his sons, though reason has dominion over the members with despotic rule, which is characteristic of a master regarding his servants."[13] On this point there is no room for contradiction. This is the case to the extent that the motion of the members is governed by the

---

13. Aristotle, *Politics* I, c. 5, 1254b4-6; cited in Thomas Aquinas, *The Summa of Theology* (*Summa theologiae*) I-II, q. 17, a. 7.

sensitive powers, as it is governed in local motion. This is clear from what was said in the preceding question.[14]

In some members motion arises much more from the disposition of the organ than from the sensitive power of the soul. Hence, those members in which such motions exist are not, with respect to those motions, naturally obedient to reason, but they are moved by an involuntary or non-voluntary motion. The Philosopher says of them at the end of *The Causes of the Motion of Animals*, "We have already said how animals are moved by voluntary motions and for what reasons. Certain of their parts are moved by some involuntary motions, as well as by many non-voluntary motions. By involuntary I mean, for example, that of the heart and that of the sex organ. For they move when something is seen; still, they are not moved at the bidding of the intellect. By non-voluntary I mean, for example, sleeping, waking and breathing and whatever others are like that. No one is master of these without qualification, nor is imagination or appetite."[15] Note that he is not speaking of the motion of the heart in pulling and pushing to cause forward motion – for that is purely voluntary and falls under the command of the will and appetite. Rather, he speaks of that which occurs in the pulse, which is purely natural and does not come from the imagination or appetite, or of that which occurs apart from the command of the will, but naturally precedes it. On this point he says in the third book of *The Soul*, "Often the intellect" thinks "of something frightful or delightful; fear or delight does not come about for that reason, but the heart is moved."[16] He also speaks of this in *The Causes of Motions*, "Imagination and intelligence have the power of real things," and so on, as in the preceding question.[17] Hence, in the same passage after what was just cited, he also assigns the cause of such motions. He says, "The causes of motions are warmth and coldness, both external and internal. The motions of the mentioned parts are natural, and they are brought about apart from reason and occur because of a change that takes place."[18] They are either brought about without imagination and intellect coming before, or they are aroused by imagination and intellect, insofar as "these are productive of emotions by introducing the forms of the things that produce them," as we mentioned in the preceding question.[19]

14. Cf. Henry of Ghent, *Quodlibet* IX, q. 5, pp. 34-45.
15. Aristotle, *The Motion of Animals* (*De motu animalium*) c. 11, 703b3-11.
16. Aristotle, *The Soul* (*De anima*) III, c. 9, 432b30-433a1.
17. Aristotle, *The Motion of Animals* (*De motu animalium* c. 7, 701b18-19; cf. Henry of Ghent, *Quodlibet* IX, q. 5, p. 57.
18. Aristotle, *The Motion of Animals* (*De motu animalium*) c. 7, 703b14-19.
19. Cf. above, p. 56.

He correctly says at the end of *The Causes of Motion*, "At times the same motion is produced in those parts in beings with intellect as is produced apart from reason, and at times not; the reason is that at times passive matter is present, at other times there is not enough matter or matter of the right quality."[20] What commands, then, should be of the proper type for the one to whom the command is given, because the latter has to obey him to carry out what was comanded. Since that which has freedom of choice is of itself bound to obey no one, but rather all the other things are bound to obey it, freedom of choice belongs of itself to the will. Hence, it belongs to other things only by participation in it, just as the power of the first mover remains in the second mover, as we have elsewhere sufficiently explained.[21] From a consideration, then, of the relation of the one commanding to him to whom the command is given, it is clear in the second place that to command is an act of the will.

Hence, away with the claim of some persons that "the root of liberty, as a cause, is reason, or intellect," though "its subject is the will," so that "the will can freely be carried to different things only because reason can have different conceptions of the good."[22] We have elsewhere shown that this is impossible. Indeed the will is both the subject and the first root of liberty; from this root it is found in the acts of reason and of the other powers, such as the virtues, through their participation in its command and through its imprint upon them, as we have elsewhere explained.[23] Hence, when the intellect precedes the will by its action, the action of the will derives its rationality from the intellect, but not its freedom of choice. When, on the contrary, the action of the will precedes the intellect, the action of the intellect has from the will freedom of choice, but not rationality. The intellect is naturally prior, and it also acts with natural priority, because we cannot will what is not known. Hence, rationality, which is a property of the intellect, is naturally prior to freedom of choice, which is a property of the will, and the will naturally has rationality from the intellect before the intellect has freedom of choice. Hence, whatever is rational has freedom of choice, and vice versa, and everything with intellect has will, and vice versa. This agrees, nonetheless, with Damascene; he says in chapter twenty-one of the first book, at the beginning, "In those beings in which there is rationality, freedom of choice immediately follows."[24] Thus rationality comes first, and freedom of choice follows, and not the reverse.

---

20. Aristotle, *The Motion of Animals* (*De motu animalium*) c. 7, 703b36-704a2.
21. Cf. Henry of Ghent, *The Ordinary Questions* (*Quaestiones ordinariae* [*Summa*]), a. 45, q. 2, 3, 4.
22. Thomas Aquinas, *The Summa of Theology* (*Summa theologiae*) I-II, q. 17, a. 1, ad 2um.
23. Henry of Ghent, *Quodlibet* I, q. 16 (*Henrici de Gandavo Quodlibet I*), ed. Macken, pp. 98-101.
24. John Damascene, *The Orthodox Faith* (*De fide orthodoxa*) c. 41, *PG* 94, 959B.

In accord with this position whereby freedom of choice is attributed to actions of the will and of reason, it ought to be understood to be essentially in the actions of the will, but by participation in the actions of the intellect. Likewise, where rationality is attributed to the actions of the will and of the intellect, it ought to be understood to be essentially in the actions of the intellect, but by participation in the actions of the will. Hence, when Damascene says in chapter twenty-four of the second book of the *Sentences*, "The will is rational, and the natural appetite is free in choice. For it is freely moved in choice and reason,"[25] I understand by "in choice" "in the will" and by "in reason," "in the intellect," so that we could likewise say it the other way around, "For it is rationally moved in reason and choice, that is, in the intellect and will." Likewise, he goes on to say, "It freely desires by choice, freely wills by choice, freely seeks by choice, freely examines by choice, freely judges by choice, freely disposes by choice, freely chooses by choice, freely makes a move by choice, freely acts and always operates by choice in those things which are in accord with its nature."[26] In these cases some actions proper to the will are attributed to reason because of the freedom of choice that is united with it. Thus, if we understand that some of them belong to the will and others to the intellect, we will understand that freedom of choice is attributed essentially to those that belong to the will and only by participation to those that belong to reason. The same thing holds the other way around, if we say conversely, "The intellect is free in choice, and the natural mind acts rationally. For it is moved rationally in reason and in free choice; hence, it desires rationally, wills rationally, seeks rationally, examines rationally, judges rationally, disposes rationally, chooses rationally, makes a move rationally, and acts and operates rationally in things in accord with its nature."

In this way both theologians and philosophers frequently interchange the intellect and the will and their activities and manners of acting, but the careful reader should distinguish each of them. Hence, when Damascene says in *The Two Natures and One Person of Christ*, "Irrational things are not freely moved to desire by choice; for when desire is naturally aroused in them, they are guided by nature, overcome by such desire, since they do not have reason which commands. Hence, they are immediately stirred to action unless they are prevented by something else, since they are not free by reason of choice, but under the power of desire."[27] Certainly, when he says, "since they do not have reason which commands," he is thinking of intellect and will together as rea-

---

25. John Damascene, *The Orthodox Faith* (*De fide orthodoxa*) c. 36, *PG* 94, 946C.
26. John Damascene, *The Orthodox Faith* (*De fide orthodoxa*) c. 36, *PG* 94, 946C.
27. Cf. John Damascene, *The Two Wills in Christ* (*De duabus in Christo voluntatibus*) c. 19, *PG* 95, 150A-C.

son. He immediately goes on clearly to distinguish them, saying: "The rational nature has the ability to yield to natural desire and the ability not to yield to it, but to conquer it."[28] He points out that the natural desire for life both exists and yields to reason. "For many using reason as their guide have willingly gone to their death and restrained their desire for food and sleep and the rest, guiding nature by free choice."[29] Notice how he distinguishes them, first saying "reason as their guide," for example, in counseling, and then saying, "guiding nature," for example, in natural appetites, "by free choice," for instance, in commanding the lower powers and restraining them from what they seek. And so, as Damascene adds after some intervening material, "As rational, man was created the king of every irrational creature; since he is both rational and intellectual, he understands and reasons in appetition; as both rational and intellective, he desires by free choice."[30]

Thirdly, it is clear that to command is an act of the will, especially from the condition of what is commanded the one to whom the command is directed by the one commanding. For what is commanded is an act to be carried out by the one who is given a command. Accordingly, no one has a command given him concerning a disposition which he himself does not have to bring about, but which he only has to have produced in him by another. Now the will can only will those things which pertain to the exercise of the other potencies, for example, what it wills the intellect to consider, to advise, and so on, or those things which the intellect determines for it in accord with what certain people say. But reason in no sense commands the first sort of willing; rather, by that willing the will commands reason, as they admit.[31] The second sort of willing is also, according to them, not caused by the will moving itself, but rather by the intellect and by the good as known, as they say. Thus the will cannot be commanded unless one takes "command" in the common sense in which members of the body are commanded with despotic command. Therefore, in no sense should we state that the intellect can command the will. But one of them necessarily is able to command the other, and it is the function of that one to command without qualification. Hence, to command without qualification is the function of the will, and not of the intellect.

---

28. Cf. John Damascene, *The Two Wills in Christ (De duabus in Christo voluntatibus)* c. 19, *PG* 95, 150C-D.

29. Cf. John Damascene, *The Two Wills in Christ (De duabus in Christo voluntatibus)* c. 7, 16, 18; *PG* 95, 135B, 143D-146A, 171B-174B.

30. John Damascene, *The Two Wills in Christ (De duabus in Christo voluntatibus,* c. 18 bis; *PG* 95, 147C.

31. Cf. Thomas Aquinas, *The Disputed Questions on Evil (Quaestiones disputatae de malo)* q. 6.

<With Regard to the Arguments>

From what has been said the objections of both sides are resolved.

Quodlibet XIV, Question 5

Are the intellect and the will equally free potencies?

With regard to the second point it was argued that the intellect and the will are equally free potencies, because the freedom of a potency lies in the fact that it can elicit its act first and by itself. But this ability belongs equally to the intellect and to the will, because it belongs to them by reason of the fact that they are spiritual powers. This is clear from what the Commentator says regarding the passage in the eighth book of the *Physics*, "It is impossible that what moves itself move itself in terms of its whole self."[1] The Commentator speaks as follows: "This is only impossible in movers that are bodies or powers in bodies."[2] Because the intellect and the will are spiritual potencies, they are not bodies or powers in bodies. For this reason, then, it is characteristic of them that it is not impossible, but rather possible for them to move themselves in terms of their whole selves. And freedom consists in this. Hence, since they are equally spiritual potencies, it follows that they are equally free.

Likewise, in the ninth book of the *Metaphysics* and in the second book of *Interpretation*, the Philosopher says that rational potencies can produce contrary effects.[3] They have this ability only because they are rational. On this basis they are distinguished from irrational and natural potencies which cannot produce contrary effects. Hence, since the intellect and the will are equally rational – or, if one is more rational, it is the intellect, because it reasons both for itself and for the will – either they are equally free, or the intellect is freer than the will.

So too, that potency is freer which in terms of its act depends less upon another and upon the act of another, because dependence upon another impedes its freedom in carrying out its act. But in its act the intellect does not depend at all on the act of the will. Rather, just the opposite is the case; in its act the will depends upon the act of the intellect. After all, the intellect is able to know without any prior act of the will, but the will is not able to will anything at all without a prior act of the intellect. Therefore, and so on.

---

1. Aristotle, *Physics* VIII, c. 5, 257b2-3.
2. Averroes, *The Commentary on the Physics* (*Phys. VIII Comm.*), ed. Junt., IV, p. 380D.
3. Aristotle, *Metaphysics* IX, c. 5, 1048a8-9 and *Interpretation* c. 13, 22b37-39.

On the contrary, that potency which freely determines itself to its act is freer than one that is necessarily determined to its act by something else. The will freely determines itself to its act, but the intellect is necessarily determined by the object. Therefore, and so on.

Here it is necessary, before all else, to see what freedom is not and what freedom is and in what it consists. In this way what I am aiming at will be clear, and clear in accord with the art of definition in which the supreme genus is supposed, for we suppose that freedom is a faculty. A faculty in itself is nothing other than an ability that is indifferent to being acted upon and to acting. But in this case, as will be seen immediately below, freedom is not found in purely passive potencies. The first difference, then, to be added to "faculty" is that it is one for acting.

Furthermore, I say that freedom is not a faculty for doing or being able to do what is good and being able to do what is bad. Nor does it consist in this. For, as Anselm says in *Free Choice*, to be able to do what is bad is not "freedom, nor part of freedom."[4] Freedom, then, is a faculty by which one can by some potency do or produce what is good. For each created being is, according to its species, ordered to some good that it has the natural ability to acquire by its own proper activity, in accord with what the Philosopher says concerning man in the second book of *The Heaven and the Earth*: "He does not have complete goodness. For something that is completely good does not lack the activity by which it is good, because the thing is that activity."[5] And the Commentator adds, "And its substance is action." Hence, "those things whose nobility lies in their action are incomplete,"[6] and to make up for that incompleteness every creature has its own action according to its proper species by which it acquires for itself the good proper to its species. This is the second difference that must be added, namely, that it is a faculty for doing what is good.

Furthermore, the potency which is able to do what is good is frequently also able not to do it, but freedom is not the ability to do this and not to do this, that is, the ability to will some good and the ability not to will it. For to the extent to which it is able not to will some good, to that extent it is subject to failure. Thus, just as, according to Anselm, "neither freedom nor a part of freedom" consists in an order to evil,[7] so too freedom does not consist in the ability not to do or not to will what is good. Hence, freedom of the will concerns a present end which it cannot fail to will. The ability to will and not to will a good does not belong to freedom without qualification, but only to free choice. Thus freedom is the faculty

---

4. Anselm, *Free Choice* (*De libero arbitrio*) I, c. 1, ed. Schmidt, I, p. 203.
5. Cf. Aristotle, *The Heaven* (*De caelo*) II, c. 12, 292b2-7.
6. Averroes, *The Commentary on the Heaven* (*Comm. in de caelo*), ed. Junt., V, p. 141K.
7. Anselm, *Free Choice* (*De libero arbitrio*) I, c. 1, ed. Schmidt, I, p. 208.

by which one can do good, and I do not mean: *not do*, but *do* good. This is the third difference.

Furthermore, according to the statement of the Philosopher in the first book of the *Metaphysics*, we say that "a man is free when he exists for his own sake, not for another's."[8] Thus only that potency is free which exists for its own sake, and one which is more of this sort is freer. But a man exists for his own sake and not for the sake of another if he does not have to care for or be concerned about someone else, but only about himself. In this sense a human lord is free in comparison to a human servant, since the servant has to be concerned about his lord and about his lord's possessions. This stems from the state of servitude which is the opposite of freedom, and not the other way round; that is, it does not stem from release from the condition of servitude which is freedom.[9] Hence, a servant is useful to his lord by a servile usefulness, by which the lord is never useful to the servant. Avicenna speaks of this in the first book of his *Metaphysics*.[10] Less commonly, there is a usefulness that draws one to someone more excellent who is like a servant to him, because he exists for his sake. But this does not hold the other way around. For the common usefulness is close to servitude,[11] but the usefulness of the more noble to the less noble is not like servitude. After all, the servant is useful to the one he serves, and the one he serves is useful to the servant. But each usefulness has its proper mode and different manner. The reason is that the usefulness by which the superior is useful to the subject is the perfection of the subject without which he cannot have being. But the usefulness by which the inferior is useful to the superior is a relationship without which the superior cannot have well-being, but might be prevented from achieving his end in a particular case. For this reason, when the Philosopher says that the wise man is by himself most self-sufficient, he adds: "But it is perhaps better to have coworkers."[12] There the Commentator says: "But this, 'perhaps' is added, because they are coworkers regarding things necessary for the body, not for happiness itself, not for that contemplation which belongs to the intellect, but for avoiding disturbance of contemplation. After all, he needs someone to cook his food, to carry water, to clean his clothes, and generally to provide for his needs. If he lacks these things, his body, either in extreme hunger or in serious illness, impedes contemplation. Still he is by himself most self-sufficient because he needs very few things. After all, he will himself draw water and cook his food. If he needed a meal of partriges or

---

8. Aristotle, *Metaphysics* I, 2, 982b25-26.
9. I read: *ex absolutione e conditione seruitutis* for *ex absolutione et conditione seruitutis*.
10. Avicenna, *Metaphysics* I, c. 3, ed. van Riet, p. 20.
11. I have dropped "*minus*" before "*communis*" in order to make better sense. The extra "*minus*" could have come from "*minus communiter*" two sentences before.
12. Cf. Aristotle, *Nicomachean Ethics* X, 7, 1177a35-1177b1.

pheasant, then he would be needy as a servant, that is, a servant of his own desires."[13] Thus, as Tully says in *Paradoxes*, "All wicked men are minds that are broken and cast down and lacking choice. But this man is truly rich. They are truly needy."[14] In the same place he also says, "Whom do we understand to be rich, or of whom is this saying true? I think it is true of him who has such great wealth that he is easily content to live generously, who seeks nothing, desires nothing, hopes for nothing more. But shall I ever call that man rich who has innumerable desires, since he himself is aware of his neediness?"[15] Thus the freedom of a man by reason of which he is called free is a faculty by which he is ordered only to himself and not to another. This is to be understood in the sense in which Avicenna speaks of free knowledge in the first book of the *Metaphysics*. "This knowledge is so lofty that it does not deign to be useful to the other sciences, though the rest are useful to it."[16] In the same way, as the Philosopher says above, that knowledge is alone free which is desired for its own sake and not for the sake of some gain. Rather, when all necessities are provided, it is sought on its own account and not on account of some other usefulness that could come to them from it.[17]

Hence, it belongs only to free men for whom all necessities are at hand. Just as a free man does not exist on account of others so that he has to be concerned about them, so he does not need others so that they have to be concerned for him as if they were useful on his account. Hence, with regard to that blessed contemplative happiness, in terms of that free knowledge which is called wisdom, by which a man also is truly free, the Philosopher says in the tenth book of the *Ethics*, "By himself he is most self-sufficient."[18] There the Commentator says: "Because he needs very few things. After all, he will himself draw water and boil his food."[19] For this reason, the Apostle spoke as one truly free in chapter nine of the First Letter to the Corinthians, "Though I was free with respect to all."[20] There the gloss says, "that is, indebted to no one."[21] He says in chapter twenty of Acts, "I have desired no one's gold and silver and clothing, because these hands supplied those things I needed."[22] For this reason

---

13. I have not been able to find this reference in Averroes.
14. Cicero, *Paradoxes of the Stoics* (*Paradoxa Stoicorum*) 35, ed. J. Molager, pp. 118-119. The last two sentences are not found in Cicero.
15. Cicero, *Paradoxes of the Stoics* (*Paradoxa Stoicorum*) 42 and 44, ed. J. Molager, pp. 124-125.
16. Avicenna, *Metaphysics* I, c. 3, ed. S. van Reit, p. 19.
17. Cf. Aristotle, *Metaphysics* I, c. 2, 982b23-29.
18. Aristotle, *Nicomachean Ethics* X, 7, 1177b1.
19. I have not found this reference.
20. 1 Co 9:19.
21. Peter Lombard, *Collectanea in epist. d. Pauli: in ep. 1 ad Cor.*, PL 191, c. 1613.
22. Cf. Ac 20:33-34.

Jerome says to Heliodorus, "Among the Greek philosophers that man is praised who boasted that everything he used was made by his own hands."[23] These points provide two more differences. Thus by adding the five differences to "faculty," which is the genus of freedom, I say that freedom is the faculty by which someone actively can do what is good for himself without needing another for this. This must be the case, if it is freedom that does the act by which one first acquires by himself his own good from a principle which is in himself. This is the fifth difference.

But this too still belongs to a purely natural potency. He says at the end of the fourth book of *Meteors*, "All things are determined."[24] It is also stated in the second book of *The Heaven and the World*, "Some of them attain goodness by one operation, and some of them attain it by many operations."[25] For this reason we must add to the description of freedom and say that it is a faculty by which one can produce good for oneself without the need of another from a principle in himself, and can do this voluntarily. This is the sixth difference. It excludes the impulse by which a purely natural potency can by its operation attain its proper good. But the intellect is also able to do this with reference to the exercise of its act since, once it has been actualized by simple understanding, it goes on to division, composition, and comparison. Thus, if nothing had to be added to the freedom in the will which was not in the freedom of the intellect, they would be both free and equally free, with the one exception that will now be stated. For it is clear that by the will the intellect can be restrained from the exercise of its act and compelled to it. By the command of the will the intellect is restrained from the activities that we have mentioned, and it is compelled to them at the pleasure of the will, as we experience in ourselves. But the will has such great freedom that it cannot be compelled to its act by anything and cannot be restrained from it by something else. Rather without any compulsion, it proceeds to its act, and without anything restraining it, it ceases from its voluntary act.

Some, as is clear from what they say, grant this regarding the will in terms of beginning its act, though they deny this in terms of stopping its act. They say that the will necessarily follows the judgment of reason. For, as they say, although it is necessarily moved to will something, freedom is still not excluded. It does not follow that, because this is necessary, it is, therefore, not voluntary or free. For the actions of active beings are found in what they act upon and affect. Thus, since the will in itself is free, it receives the movement of willing according to the manner of its nature, namely, freely, though necessarily. They appeal to the example in the servant who, insofar as he moves at the order of his master, in a sense

---

23. Cf. Jerome, *Letter* (*Epistola*) LX, 12: *PL* XXII, 597.
24. Aristotle, *Meteors* (*Meteorologica*) IV, c. 12, 390a10-11.
25. Aristotle, *The Heaven* (*De caelo*) II, c. 12, 292b10-10.

moves necessarily and yet freely, because he is pleased to carry out the order. In that way there is in the will the character of mover and the character of moved in the same simple reality, and in that way it moves itself. Yet it necessarily follows the judgment of reason or of the intellect which weighs the matter in its principles and freely judges by itself. From this freedom, as they say, there is derived the freedom in the will, and thus the will is moved in a free manner according to reason itself. In accord with this, they say that, if the will were not moved according to reason, it would not be moved freely. Thus freedom of the will is entirely derivative from the freedom of the intellect. They also say that the intellect is, on these grounds, freer than the will.

But if this is the way it is with the motion of the will regarding the beginning and start of the act of willing by the will, there is no difference in the motion of the will toward the end and in those things which are means toward the end. In fact, just as by the judgment of reason it is moved toward a present end freely, though not by free choice, it is likewise moved in that way to those things which are means to the end. Although freedom of the will can be preserved along with the necessity of its motion, freedom of choice cannot in any sense be preserved, because the will of free choice can neither begin the act of willing nor stop it against the judgment of reason. We constantly experience the opposite of this in our own case with regard to those things which are means to an end. But this has been sufficiently discussed elsewhere.[26] Hence, however much reason ponders and judges, this does not at all pertain to the free choice of the will, nor, consequently, to its freedom, because freedom and free choice always follow the same course with regard to those things which are means to an end. In that way, although the intellect has nothing from the motion of the will regarding the beginning of its act, because the intellect can begin its act without a previous motion of the will, still the intellect has much from the will regarding the exercise of its act in beginning and stopping it, because it can neither begin nor exercise it against the command of the will. In this way the intellect falls far short of the freedom of the will, because the intellect is not moved to the exercise of its act unless it is first moved to the act of simple intelligence through the determination of the first act of understanding by the object, as has often been said and is now repeated. But the will moves to first act without being moved or determined by something else, and also to every exercise of its act and to the command of the acts of the other powers. Therefore, I state without qualification that, though the intellect is free in some sense, as will now be clear, the will is much freer than the intellect. Thus in the definition of freedom which is proper to the will, it is necessary to

---

26. Cf. above, pp. 50-64.

add something over and above the freedom of the intellect, namely, the seventh condition, and say that freedom of the will is the faculty by which it is able to proceed to its act by which it acquires its good from a spontaneous principle in itself and without any impulse or interference from anything else.

In this respect the intellect falls short, because through the determination of its act to the first act of understanding, that is, of simple intelligence, it is compelled by the object; moreover, by its freedom the will compels or does not compel the intellect to the exercise of its act, and it freely restrains it from it. From these points it is clear that the intellect has freedom only for the exercise of its act, as long as the will permits. For this reason I state without qualification that the will is much freer than the intellect. The last reason adduced for this should be conceded, and I think that the Philosopher understood this sort of freedom with regard to the will when he said in the second book of the *Ethics* that it is "the master of its actions."[27] Such freedom of the will is so great with regard to means to an end that it moves toward them not only freely, but also by free choice, while the intellect, even if it can do something freely, still cannot do anything with free choice. Hence, Tully said in *Paradoxes*, "What is freedom? The power of living as you will. Who lives as he wills save he who wills rightly, who does not obey the laws out of fear, who says nothing, does nothing, thinks nothing save willingly and freely, whose every counsel and undertaking arise from him, and in whose possession there is nothing more valued than his will and judgment? Thus the wise man alone does nothing against his will, nothing in sorrow, nothing under coercion. Hence, we must admit that only one so disposed is free. Or, is he free who can refuse nothing to one who commands him?"[28]

<With Regard to the Arguments>

To the first opposing argument, that the will and the intellect of themselves elicit their acts equally first and equally by themselves, I say that it is not true. What they add in proof of this, namely, that this belongs to a potency by reason of the fact that it is spiritual, I grant is true. But, if it is impossible that something be moved by itself only in movers that are bodies or powers in a body, the opposite need not follow, namely, that it is equally possible for all powers that are not in a body to move

---

27. This citation is not found in the second book of Aristotle's *Nicomachaean Ethics*. The phrase is elsewhere attributed to John Damascene in *The Orthodox Faith (De fide orthodoxa)* c. 41, *PG* 94, 964A; cf. *Henrici de Gandavo Quodlibet IX*, q. 5, ed. Macken, p. 130, l. 5; cf. above, p. 56.
28. Cf. Cicero, *The Paradoxes of the Stoics (Paradoxa Stoicorum)* 34-36, ed. J. Molager, pp. 118-119.

themselves. Nor is it necessary as well that it be equally possible[29] for all movers which are not bodies, for angels move themselves more freely than men both in terms of the will and in terms of the intellect.

Still it is strange that some try to claim that the Commentator is there speaking only of separated spiritual beings in which the intellect and its object are identical, because he mentions them at the end of the comment. In fact, he is not speaking only of them and intends merely to make them an exception from what Aristotle says; rather, he points out an error that he attributes to Plato with regard to these spiritual realities, as is clear from examining and weighing his text which is as follows: "This is impossible only in movers that are bodies or powers in bodies."[30] Nothing more than that in the words of the Commentator pertains to establishing the truth of the statement of Aristotle that was mentioned. The remaining things which he adds do not pertain to an explanation of the statement of Aristotle that we mentioned; rather, in these things he turns from what he said in his brief explanation to certain claims of Plato with which he agrees in part and disagrees in part. He first draws a conclusion from his explanation in which he harmonises Plato and his own explanation. He says: "Thus Plato thought that things moved by themselves are moved by abstract movers," and so on, "because, of course, he held that something moving itself is an abstract mover and that an abstract mover is something moving itself."[31] For as the Commentator says shortly before his explanation already mentioned, "Plato thought that something moved by itself is composed of what is moved and a mover that moves itself."[32] This is, of course, true, as the Commentator says, of something moved and a mover that is not moved by another at all, as the heaven is moved and as man is moved by the soul. For the heaven is not moved by a mover, nor is man in his body moved by the soul, unless the moved thing first moves itself by the motion of the will. But that brute animals do not move themselves in terms of place is due only to the fact that they are first moved by another in terms of appetite. Hence, he goes on in his comment and establishes the statement of Plato's that he had already set forth: "And thus Plato thought," etc., adding: "But if this were the case, then the mover and the moved are in that case spoken of equivocally with the motion that is found on earth."[33] The motion found there, that is, in abstract things, where they move themselves first by will, and the motion found on earth, namely, in bodies moved by those things, for instance, the heavenly and the human body, are truly spoken of with a great deal of

29. I have changed the Latin which reads: "impossible."
30. Averroes, *The Commentary on the Physics* (*Phys. VIII Comm.*), ed. Junt., p. 380E.
31. Averroes, *The Commentary on the Physics* (*Phys. VIII Comm.*), ed. Junt., p. 380DE.
32. Averroes, *The Commentary on the Physics* (*Phys. VIII Comm.*), ed. Junt., p. 380B.
33. Averroes, *The Commentary on the Physics* (*Phys. VIII Comm.*), ed. Junt., p. 380E.

equivocation. This is clear from what has already been said above in the second question on the motion of the will from itself by the motion of volition and the motion of what is heated from the motion of that which heats it. But since Plato did not see this equivocation, as the Commentator attributes it to him, he immediately finds fault with him on this point and adds: "And this error befell Plato as a result."[34] But even before he explains the point on which he erred, he sets forth a point on which he commends him, saying that "he thought that the first mover in a body which moves of itself is not a body, and the idea is correct."[35] Then he immediately adds the point on which he finds fault with him and says, "And he thought along with this that every mover, whether it is a body or not, only moves if it is moved by one motion understood univocally,"[36] and this point is certainly false, as we have explained in the preceding second question. Nonetheless, Aristotle also attributes this view about the soul to Plato, when he argues against him in the first book of *The Soul* about Plato's claim that the soul moves itself.[37] He brings in arguments against true bodily motion as if Plato had claimed that the soul is moved by itself with a motion univocally the same as that by which the body is moved by the soul.[38] At the end of his book Macrobius replies to the arguments of Aristotle with the claim that motion is used equivocally in the case of the soul's motion by itself and the motion of the body.[39] He says that it was not Plato's idea that the soul moved itself with univocally the same sort of motion as that by which the body is moved.[40] In this way one must reply to all the arguments taken from the statements of Aristotle and others concerning the motions of bodies to prove that the soul or the will cannot move itself. Hence, Aristotle says in the eighth book of the *Physics*, prior to the proposition mentioned, "How is it possible that something continuous by nature move itself?" Rather, "it is necessary that the mover in each of them be distinct from the moved."[41] As the Commentator says, "It must be distinct from the moved either according to both definition and being, as those things which are moved from the outside, or according to definition only, as is the disposition in things having souls. For the soul which is the mover in them is distinct from the body that is moved according to definition, although it is not distinguished according to being.

---

34. Averroes, *The Commentary on the Physics* (*Phys. VIII Comm.*), ed. Junt., p. 380E.
35. Averroes, *The Commentary on the Physics* (*Phys. VIII Comm.*), ed. Junt., p. 380E.
36. Averroes, *The Commentary on the Physics* (*Phys. VIII Comm.*), ed. Junt., p. 380E.
37. Cf. Aristotle, *The Soul* (*De anima*) I, c. 3, 406b15-16.
38. Cf. Aristotle, *The Soul* (*De anima*) I, c. 3, 406b26ff.
39. Cf. Macrobius, *The Commentary on the Dream of Scipio* (*Comm. in somnium Scipionis*) 2.15-16, ed. J. Willis (Leipzing: Teubner, 1970), pp. 140-151.
40. Cf. Macrobius, *The Commentary on the Dream of Scipio* (*Comm. in somnium Scipionis*) 2.15.18-16.7, ed. J. Willis, pp. 143-147.
41. Cf. Aristotle, *Physics* VIII, c. 4, 255a13,17.

For it is impossible that the soul be without the body except in an equivocal sense."[42] If then someone would want to apply this statement of Aristotle and the Commentator to spiritual things with regard to the motion of their will, one should say, in agreement with what the Commentator says here, that it is not impossible in spiritual beings, but that Aristotle is speaking only of the motion of bodies, as the examples of the Commentator show as well. Now motion is predicated equivocally of bodily and spiritual motion, as has been said. Next the Commentator repeats the statement in which he had commended Plato in order to add what he also attributes to him. He says, "And it was impossible in his view that a mover that is a body move itself, and it is likewise impossible that a mover that is a power in a body do so. Then Plato concludes from this that the soul is not in a body and that it is eternal, since it moves itself."[43] One must distinguish what the Commentator says is the case in what he immediately adds. "And this would be true if the soul moved itself essentially."[44] He correctly indicated this distinction that something moving itself moves itself either essentially or not essentially, when he says that this statement, namely, that soul is not in the body, "would be true, if the soul moved itself essentially."[45] He implies that it would not be true if the soul moved itself non-essentially and that it could, then, very well both be in the body and move itself. And to explain what it means to move oneself essentially, he adds: "And it would move itself with the motion proper to abstract things, namely, so that the intellect and its object would be identical in it, as was said about the first mover and the other abstract movers."[46] By this he intended, on the contrary, that it moved itself by a non-essential movement not proper to abstract things, namely, so that the intellect and its object are not identical. Thus he understands that only those things are moved by themselves essentially in which there is an identity of the intellect and its object and that those things do not move themselves essentially in which there is not an identity of the intellect and its object. Thus he understands that what moves itself in the first way is only something abstract and is not in a body or a power in a body, as are, according to the Philosopher, all the separated movers. But what moves itself in the second way is not abstract, but is a power in a body, although it is not educed from the potency of a body as a material form, as are, according to the Philosopher, the soul and the conjoined mover of the heaven. Hence, as has already been said, according to the Commentator, "it is impossible that the soul be without the body except in an equivocal

42. Averroes, *The Commentary on the Physics* (*Phys. VIII Comm.*), ed. Junt., p. 367F.
43. Averroes, *The Commentary on the Physics* (*Phys. VIII Comm.*), ed. Junt., p. 380EF.
44. Averroes, *The Commentary on the Physics* (*Phys. VIII Comm.*), ed. Junt., p. 380F.
45. Averroes, *The Commentary on the Physics* (*Phys. VIII Comm.*), ed. Junt., p. 380F.
46. Averroes, *The Commentary on the Physics* (*Phys. VIII Comm.*), ed. Junt., p. 380F.
47. Averroes, *The Commentary on the Physics* (*Phys. VIII Comm.*), ed. Junt., p. 367F.

sense,"[47] and he holds the same position regarding the mover of the heaven. From this it is clear that his statement about things that move themselves in which there is an identity of intellect and its object does not belong to his explanation of the proposition of Aristotle's that has been mentioned: "It is impossible that what moves itself moves itself in terms of its whole self."[48] Nor can it be said that in explaining this proposition the Commentator intended by the words, "This is not impossible except in movers which are bodies or powers in a body,"[49] that it is only possible in spiritual things in which there is an identity of the intellect and its object. Indeed he clearly implies according to what has already been said that all spiritual beings in which the intellect, its knowing and its object are not the same, and similarly the will, willing and the object willed, could move themselves non-essentially, if there should be such beings. Although, according to him and according to the Philosopher, there were no such beings, all spiritual beings apart from the one God are such. He implies that it is not impossible that some of them be an act of a body, for example, the soul, and still not be a bodily power that is educed from the potency of matter. And the same is true of the conjoined mover of the heaven. He clearly implied this too above where he said: "The heaven is not moved by the mover of the heaven," and so on.[50] Thus the Commentator understands that this is possible in all spiritual beings, of which motion is predicated equivocally in comparison with the motion of a body. They agree on this point. For this reason the Commentator added at the end of the comment mentioned, "But this motion is spoken of equivocally with the motion that is from bodies or from movers that are bodies or powers in bodies."[51]

From what has been said in the second question it is clear that motion is predicated equivocally not only of that in which the intellect and its object, or the will and the willed, are identical, and of bodies. It is also predicated equivocally of the will in which the will, the willing and the willed are distinct, and of bodies, although, as has been said, motion is also predicated far more equivocally of the motion of the will in the soul and of the motion of God. Hence, the motion of bodies and of the will both agree on this point, but they differ from the motion of God, because in God there is a real identity of mover, moved, and motion, while in the will there is an identity of the mover and the moved, from which the motion is really different. In bodies, however, the mover and the moved really differ, and the motion is likewise different from both of them.

------

48. Aristotle, *Physics* VIII, c. 5, 257b2.
49. Averroes, *The Commentary on the Physics* (*Phys. VIII Comm.*), ed. Junt., p. 380D.
50. I have not been able to find this reference.
51. Averroes, *The Commentary on the Physics* (*Phys. VIII Comm.*), ed. Junt., p. 380F.

To the second objection that rational potencies can produce contrary effects by reason of the fact that they are rational, I say that it is true when they are purely active potencies and insofar as they are active. For where Aristotle says this, he is speaking only of the comparison of active rational potencies and non-rational, but natural ones. To be able to produce contrary effects is an indication of freedom only in active potencies. For however much passive potencies are able to produce contrary effects, still freedom cannot for this reason belong to them, because it could only belong to them in relation to being acted upon, whereas freedom belongs to active potencies only in relation to acting. But everything acted upon insofar as it is acted upon is acted upon by a natural impulse, and it could not fail to be acted upon when the active principle is present, whether it acts naturally or by free choice; this, however, is incompatible with freedom. Hence, it is very strange to say that the will is a passive power and still is free, just as it would be strange to say that something is an active natural power and still is free. For this reason it happens that the will is found to do one of two equally contingent actions, but a natural agent cannot, as the Commentator says on the second book of the *Physics*, "It is impossible that from something equally contingent there arise one of two actions except as the result of another cause joined with it, because neither action is more worthy," and so on.[52] Since this action occurs, it does not arise from what is equally contingent unless another extrinsic cause is joined with it. But from what is equally contingent insofar as it is equally contingent, no action arises, because its nature is the nature of matter and not the nature of form. He says the same thing about the soul in the second book.[53] No active potency can be indifferent to both of two contraries. In fact, every potency of this sort is passive and behaves like matter, in which one of those contraries is reduced to second act by another agent, whether this potency is genuine matter or the soul.[54]

For this reason it is also strange that some raise this point to prove that the will is a potency able to produce contraries. For we say to this, as we did above, that this is only impossible in movers which are bodies or powers in bodies, since they do not act save naturally. But in spiritual things which act freely this is by no means impossible, and thus in bodily things to be equally indifferent to contraries is merely characteristic of matter and of what is passive. But in spiritual things it can quite well be chararcteristic of form and of act. Although the intellect, then, is equally rational or even more rational than the will, it is not an active

52. Averroes, *The Commentary on the Physics* (*Phys. VIII Comm.*), ed. Junt., p. 66IK.
53. "*Idem supra secundum de anima*" would more naturally mean: "He says the same thing in the second book of *The Soul*." But the reference seems to be to Averroes' commentary on the *Physics*; cf. the previous and the following reference.
54. Cf. Averroes, *The Commentary on the Physics* (*Phys. VIII Comm.*), ed. Junt., p. 67CD.

potency, unless it is first acted upon. It is not active in terms of the exercise of an act even in judging – which is the activity that it performs more freely – without the agreement of the will, as has been already stated. And thus the intellect cannot produce contraries as freely as the will.

But they say that, if the will is moved freely, this is due to the fact that it is moved according to reason, because, if it were not moved according to reason, it would not be moved freely. I say that "reason" is the name of a spiritual potency distinct from bodily substances; it includes in common both the appetitive potency which is the will and the cognitive potency which is the intellect, although in the usage of the term, "reason" has been appropriated by the intellect. Intellect and will are distinct from the appetitive and cognitive potencies in bodily beings and from reason taken in the sense common to both. In that sense both intellect and will are called rational, and the intellect is no more rational than the will. For, as the intellect rationally or reasonably knows for itself and for the will, so the will rationally wills for itself and for the intellect. But in the sense of "reason" in which the intellect is properly called reason, the will is no more called rational that the intellect is called volitional. For as the intellect is not called volitional because it follows the command of the will which it cannot fail to follow, so the will is not called rational because it follows the judgment of reason, especially since it need not follow it. If it is in some way called rational after reason, this is accidental, because it chooses to follow reason when it need not follow it. Nonetheless, the contrary does not follow, namely, that it is accidental that, because the intellect follows the command of the will, it is called volitional, since it cannot fail to follow it. Hence, when the will is moved toward a present end, in as much as it cannot but be moved toward it, it is said to be moved toward it naturally rather than rationally, though it is moved freely. The will is in this way not called rational in the proper sense except because by the choosing of free choice it follows the correct judgment of reason as a necessary condition. For this reason some persons seem to me to be deficient when they say that the freedom of the will lies only in the freedom of reason, because reason is part of the definition of the will and, as a result, there is freedom in the will derived from the freedom of reason by which it judges freely. And the will is moved freely in accord with reason. But it is false that reason is part of the definition of the will, unless one takes reason in the wide sense as containing intellect and will.

If someone says that it is at least part of the definition of free will, I say that this too is false, except accidentally. Insofar as the will follows the correct judgment of reason, the will can be called rational. In the same way, freedom can be called rational from reason in the sense in which it is proper to the intellect, as was said. Still freedom does not, on this account, come to be in the will either from reason or from the freedom of its judgment. And so, the will is not said to be moved freely in accord with reason because its freedom is essentially derived from reason, but because through the choosing of the will it happens to follow the correct judgment of reason. For this reason alone, free choice is defined as a faculty of will and reason.

To the third argument that the intellect is freer, because in its act it is less dependent upon the will than the will is dependent in its act upon the intellect, I say this: though this is true with reference to the determination of its acts insofar as the will requires a prior act if it is to be determined to its own act and the converse is not so, as the objection would have it, the intellect is much more dependent with reference to the determination of its act upon the object of the will than the will is dependent upon the intellect. After all, the intellect not only requires the presence of its object so that, when it is present, it can bring itself to act without undergoing anything from the object in the way that the will requires the presence of the act of the intellect, but the intellect also has to be acted upon by its object before it can elicit any act of knowing. Yet, in no way does the will first have to be acted upon by the intellect in terms of its act. We have stated this point elsewhere quite often, explaining how the will is dependent upon the intellect only as upon a cause without which it cannot act. But the intellect is dependent upon its object as upon a cause on account of which it acts in the way it does, and this is true for the speculative intellect. In a special way the practical intellect has the determination of its act only from the motion of the will, because the good is not the object of the practical intellect insofar as it is good without qualification, but only insofar as it is the willed good. Otherwise, in whatever way the intellect knows the good even in its character as good, it knows it only as the speculative intellect. But if the speculative intellect first knows the good in its character as good without qualification and then later knows it in its character as the willed good, on account of which it first has the character of something doable, then the intellect that was at first speculative has become practical by a certain extension.

Bibliography

Abbreviations Used in the Bibliography:

CC   *Corpus Christianorum*
CSEL   *Corpus Scriptorum Ecclesiasticorum Latinorum*
PG   *Patrologia Graeca*
PL   *Patrologia Latina*

Primary Sources

Algazel. *Algazel's Metaphysics, a Mediaeval Translation*. Ed. J. T. Muckle. Toronto, 1933.

Anselm of Canterbury. *Cur Deus homo*. In *S. Anselmi Cantuariensis Archiepiscopi Opera Omnia*. 5 vols. Ed. F. S. Schmitt. Edinburgh: Thomas Nelson, 1946. Vol. 2.

_____. *De libero arbitrio*. In *S. Anselmi Cantuariensis Archiepiscopi Opera omnia*. 5 vols. Ed. F. S. Schmitt. Edinburgh: Thomas Nelson, 1946. Vol. 2.

Aristotle. *Analytica priora et posteriora*. Ed. W. D. Ross. Oxford: Clarendon, 1964.

_____. *Aristotle's Metaphysics. A Revised Text with Introduction and Commentary*. 2 vols. Ed. W. D. Ross. Oxford: Clarendon: 1924.

_____. *The Basic Writings of Aristotle*. Ed. Richard McKeon. New York: Random House, 1941.

_____. *De anima*. Ed. W. D. Ross. Oxford: Oxford University Press, 1956.

_____. *De animalium incessu*. In *Aristotle: Parts of Animals, Movement of Animals, Progression of Animals*. Cambridge, MA: Harvard University Press, 1945.

_____. *De caelo*. Ed. J. L. Stocks. Oxford: Clarendon, 1922.

_____. *Ethica Eudemia*. Ed. R. R. Walzer and J. M. Mingay. Oxford: Clarendon, 1991.

_____. *Ethica Nicomachea*. Ed. W. D. Ross. Oxford: Clarendon, 1925.

_____. *De motu animalium*. In *Aristotle: Parts of Animals, Movement of Animals, Progression of Animals*. Cambridge, MA: Harvard University Press, 1945.

_____. *Opera*. Ed. I. Bekker. 5 vols. Berlin, 1831-1870.

_____. *Parva naturalia*. Ed. D. Ross. Oxford: Clarendon, 1955.

_____. *Physica*. Ed. W. D. Ross. Oxford: Clarendon, 1960.

_____. *Politica*. Ed. D. Ross. Oxford: Clarendon, 1964.

_____. *Topica et Sophistici Elenchi*. Ed. D. Ross. Oxford: Clarendon, 1963.

Augustine of Hippo. *Confessiones*. Ed. L. Verheijen. *CC* 27.

_____. *Contra Faustum*. Ed. J. Zycha. *CSEL* 25.

_____. *De doctrina christiana*. Ed. J. Martin. *CC* 32.

_____. *De Genesi ad litteram*. Ed. J. Zycha. *CSEL* 28.

_____. *De libero arbitrio*. Ed. W. M. Green. *CC* 29.

_____. *De trinitate*. Ed. W. J. Mountain. *CC* 50 and 50/A.

_____. *Enarrationes in Psalmos*. Eds. D. Dekkers and J. Fraipont *CC* 38-40.

Averroes. *Commentarium magnum in Aristotelis De anima libros. Corpus Commentariorum Averrois in Aristotelem, Versionum Latinarum* 6/1. Ed. E. F. Stuart Crawford. Cambridge, MA, 1953.

_____. *Opera omnia*. In *Aristotelis opera cum Averrois commentariis*. 12 vols. Venetiis: Apud Junctas, 1562-1574; repr. Frankfurt a. M.: Minerva, 1962.

Avicenna. *Liber de philosophia prima sive scientia divina*. *Avicenna Latinus*. 3 vols. Ed. Simone Van Riet. Louvain-La-Neuve: E. Peeters; Leiden: E. J. Brill, 1977, 1980 and 1983.

Cicero, Marcus Tullius. *Paradoxa Stoicorum*. In *Les paradoxes des Stoiciens*. Ed. Jean Molager. Paris: Belles Lettres, 1971.

Henry of Ghent. *Lectura ordinaria super Sacram Scripturam Henrico de Gandavo adscripta*. *Henrici de Gandavo Opera Omnia XXXIV*. Ed. Raymond Macken. Leuven: University Press; Leiden: E. J. Brill, 1980.

_____. *Quodlibet I*. *Henrici de Gandavo Opera Omnia V*. Ed. Raymond Macken. Leuven: University Press, 1979.

_____. *Quodlibet II*. *Henrici de Gandavo Opera Omnia V*. Ed. Robert Wielockx. Leuven: University Press, 1983.

_____. *Quodlibet VI*. *Henrici de Gandavo Opera Omnia X*.Ed. G. A. Wilson. Leuven: University Press, 1987.

_____. *Quodlibet VII*. *Henrici de Gandavo Opera Omnia XI*. Ed. G. A. Wilson. Leuven: University Press, 1991.

_____. *Quodlibet IX*. *Henrici de Gandavo Opera Omnia XIII*. Ed. Raymond Macken. Leuven: University Press, 1983.

_____. *Quodlibet X*. *Henrici de Gandavo Opera Omnia XIV*. Ed. Raymond Macken. Leuven: University Press, 1981.

_____. *Quodlibet XII*. *Quaestiones 1-30*. *Henrici de Gandavo Opera Omnia XVI*. Ed. J. Decorte. Leuven: University Press, 1987.

_____. *Quodlibet XII*. *Tractatus super facto praelatorum et fratrum*. *Henrici de Gandavo Opera Omnia XVII*. Eds. L. Hödl and M. Haverals. Leuven: University Press, 1989.

_____. *Quodlibet XIII*. *Henrici de Gandavo Opera Omnia XVIII*. Ed. J. Decorte. Leuven: University Press, 1987.

_____. *Quodlibeta*. 2 vols. Paris, 1518; repr. Leuven, 1961.

_____. *Summa quaestionum ordinariarum*. 2 vols. Paris, 1520; repr. St. Bonaventure, NY, 1953.

Jerome. *Epistula LX. PL* 22.

John Damascene. *De duabus in Christo voluntatibus. PG* 95.

_____. *De fide orthodoxa. PG* 94.

Macrobius, Ambrosius. *Commentarii in Somnium Scipionis.* In *Macrobius.* 2 vols. Ed. J. Willis. Vol. 2. Leipzig: Teubner, 1970.

Peter Lombard. *Collectanea in epistulas divi Pauli. PL* 191.

Pseudo-Anselm. *De similitudinibus. PL* 159.

Pseudo-Dionysius. *De divinis nominibus. PG* 3.

Themistius. *Commentaire sur le traité de l'âme d'Aristote.* Tr. Guillaume Moerbeke; ed. Gerard Verbeke. Louvain: Publications universitaires, 1957.

Thomas Aquinas. *Quaestiones disputatae.* Ed. R. Spiazzi. Rome: Marietti, 1953.

_____. *Summa theologiae.* In *S. Thomae Aquinatis Opera Omnia.* Rome: Leonine edition, 1888-1904. Vols. 4-12.

_____. *Basic Writings of Saint Thomas Aquinas.* 2 vols. Ed. Anton C. Pegis. New York: Random House, 1945.

Secondary Sources

Auer, Johann. *Die Entwicklung der Gnadenlehre in der Hochscholastik.* Freiburg: Herder, 1951.

Brady, Ignatius. "St. Bonaventure's Doctrine of Illumination: Reactions Medieval and Modern." *Southwestern Journal of Philosophy* 5 (1974), 27-37.

Brown, Jerome V. "Abstraction and the Object of the Human Intellect according to Henry of Ghent." *Vivarium* 11 (1973), 80-104.

_____. "Henry of Ghent on Internal Sensation." *Journal of the History of Philosophy* 10 (1972), 15-28.

_____. "Intellect and Knowing in Henry of Ghent." *Tijdschrift voor Filosofie* 37 (1975), 490-512 and 692-710.

_____. "John Duns Scotus on Henry of Ghent's Arguments for Divine Illumination." *Vivarium* 14 (1976), 94-113.

_____. "John Duns Scotus on Henry of Ghent's Theory of Knowledge." *The Modern Schoolman* 56 (1978), 1-29.

_____. "The Meaning of 'Notitia' in Henry of Ghent." In *Sprache und Erkenntnis im Mittelalter.* 2 vols. Ed. J. Beckmann, Berlin: de Gruyter, 1981. Vol. 2, 992-998.

_____. "Sensation in Henry of Ghent: A Late Mediaeval Aristotelian Augustinian Synthesis." *Archiv für Geschichte der Philosophie* 53 (1971), 238-266.

Brown, Stephen. "Avicenna and the Unity of the Concept of Being." *Franciscan Studies* 25 (1965), 117-150.

Burrell, David. "John Duns Scotus: The Univocity of Analogous Terms." *The Monist* 49 (1965), 639-658.

Cunningham, F. A. "Some Presuppositions in Henry of Ghent." *Pensamiento* 25 (1969), 103-143.

De Wulf, Maurice. *Etudes sur Henri de Gand.* Bruxelles, 1895.

_____. *Histoire de la philosophie scolastique dans les Pays-Bas et la principauté de Liége jusqu'à la Révolution française.* Bruxelles, 1894-1895.

_____. *History of Medieval Philosophy.* 2 vols. 3rd ed. London: Longmans, Green, 1938.

Ehrle, Franz. "Beiträge zu den Biographien berühmter Scholastiker: Heinrich von Gent." *Archiv für Literatur- und Kirchengeschichte des Mittelalters* I (1885), 365-401 and 507-508.

Fairweather, E. "Henry of Ghent." In *The Encyclopedia of Philosophy*. 8 vols. New York: Macmillan, 1967. Vol. 4, 475-476.

Gilson, Etienne. *History of Christian Philosophy in the Middle Ages*. New York: Random House, 1955.

Gómez Caffarena, José. "Cronología de la Suma de Enrique de Gante por relación a sus 'Quodlibetos.'" *Gregorianum* 38 (1957), 116-133.

_____. *Ser Participado y Ser Subsistente en la Metafísica de Enrique de Gante*. Rome: Gregorian University Press, 1958.

Grabmann, Martin. "Berhard von Auvergne, O.P. (d. nach 1304), ein Interpret und Verteidiger des hl. Thomas von Aquin aus alter Zeit." In *Mittelalterliches Geistesbelen*. 2 vols. Munich: M. Hüber, 1936. Vol. 2, 547-558.

Hocedez, Edgar. "Giles de Rome et Henri de Gand sur la distinction réelle." *Gregorianum* 8 (1927), 358-394.

Hoeres, Walter. "Wesen und Dasein bei Heinrich von Gent und Duns Scotus." *Franziskanische Studien* 2-4 (1965), 121-186.

Hissette, Roland. *Enquête sur les 219 articles condamnés à Paris le 7 mars 1277*. Louvain: Publications Universitaires; Paris: Vander-Oyez, 1977.

Hyman, Arthur, and James Walsh (eds). *Philosophy in the Middle Ages*. 2nd ed. Indianapolis: Hackett, 1983.

Kelley, Francis E. "Two English Thomists: Thomas Sutton and Robert Orford vs. Henry of Ghent." *The Thomist* 45 (1981), 345-387.

Klocker, Harry R. "Two 'Quodlibets' on Essence/Existence." *The Thomist* 46 (1982), 267-282.

Laarman, Matthias. "Bibliographia auxiliaris de vita, operibus et doctrina Henrici de Gandavo." *Franziskanische Studien* 73 (1991), 324-366.

Lottin, Odon. *Pyschologie et morale aux XII et XIII siècles*. 2 vols. Gembloux: J. Duculot, 1954.

Macken, Raymond. "Der Aufbau eines wissenschaftlichens Unternehmens: Die 'Opera Omnia' des Heinrich von Gent." *Franziskanische Studien* 65 (1983), 82-96.

_____. "Les corrections d'Henri de Gand à sa Somme." *Recherches de théologie ancienne et médiévale* 44 (1977), 55-100.

_____. "Les corrections d'Henri de Gand à ses Quodlibets." *Recherches de théologie ancienne et médiévale* 40 (1973), 5-51.

_____. "Les diverses applications de la distinction intentionelle chez Henri de Gand." In *Sprache und Erkenntnis im Mittelalter*. 2 vols. Ed. J. Beckmann. Berlin: de Gruyter, 1981. Vol. 2, 769-776.

_____. "La doctrine de S. Thomas concernant la volonté et les critiques d'Henri de Gand." In *Tommaso d'Aquino nella storia del pensiero*. Atti del Congresso Internazionale Roma-Napoli, 1974. Napoli, 1976. Pp. 84-91.

_____. "Der geschaffene Wille als selbstbewegendes geistiges Vermögen in der Philosophie des Heinrich von Gent." In *Historia Philosophiae Medii Aevi*. Studien zur Geschichte der Philosophie des Mittelalters. Festschrift für Kurt Flasch zum 60. Geburtstag. Ed. B. Moysisch and O. Pluta. Amsterdam: Grüner, 1991. Pp. 561-572.

_____. "Heinrich von Gent im Gespräch mit seinen Zeitgenossen über die menschliche Freiheit." *Franziskanische Studien* 59 (1977), 125-182.

_____. "Henri de Gand." In *Dictionnaire d'histoire et de géographie ecclésiastiques*. Paris: Letouzey et Ané, 1990. Vol. 23, cc. 1133-1136.

_____. "L'interpénétration de l'intelligence et de la volonté dans la philosophie d'Henri de Gand." In *L'homme et son univers au moyen âge*. Actes du Septième Congrès International de Philosophie Médiévale. Ed. C. Wénin. Louvain-La-Neuve: Editions de l'Institut Supérieur de Philosophie, 1986. Pp. 808-814.

_____. "La liberté humaine dans la philosophie d'Henri de Gand." In *Regnum Hominis et Regnum Dei*. Acta Quarti Congressus Scotistici Internationalis, Padova, 1976. Studia Scholastico-Scotista, 6. Rome, 1978. Vol. 1, 577-584.

_____. "La personnalité, le caractère et les méthodes de travail d'Henri de Gand." *Theologische Zeitschrift* 45 (1989), 192-206.

_____. "De radicale tijdelijkheid van het schepsel volgens Hendrik van Gent." *Tijdschrift voor Filosofie* 31 (1969), 490-581.

_____. "Selbstverwirklichung in der Anthropologie des Heinrich von Gent." In *Renovatio et Reformatio: wider das Bild vom 'finsteren' Mittelalter. Festschrift für Ludwig Hodl zum 60. Geburtstag.* Eds. M. Gerwing and G. Ruppert. Münster: Aschendorff, 1985. Pp. 131-140.

_____. "Les sources d'Henri de Gand." *Revue philosophique de Louvain* 76 (1978), 5-28.

_____. "La théorie de l'illumination divine dans la philosophie d'Henri de Gand." *Recherches de théologie ancienne et médiévale* 39 (1972), 82-112.

_____. "La volonté humaine, faculté plus élevée que l'intelligence selon Henri de Gand." *Recherches de Théologie ancienne et médiévale* 42 (1975), 5-51.

Marenbon, John. *Later Medieval Philosophy: (1150-1350) An Introduction.* London: Routledge and Kegan Paul, 1987.

Marrone, Steven P. "Henry of Ghent and Duns Scotus on the Knowledge of Being." *Speculum* 63 (1988), 22-57.

_____. "Matthew of Aquasparta, Henry of Ghent and Augustinian Epistemology after Bonaventure." *Franziskanische Studien* 65 (1983), 252-290.

_____. *Truth and Scientific Knowledge in the Thought of Henry of Ghent.* Cambridge, MA: The Medieval Academy of America, 1985.

Maurer, Armand. "Henry of Ghent and the Unity of Man." *Mediaeval Studies* 10 (1948), 1-20.

Owens, Joseph. *An Interpretation of Existence.* Milwaukee: Bruce, 1968.

Paulus, Jean. *Henri de Gand. Essai sur les tendances de sa métaphysique.* Paris: J. Vrin, 1938.

_____. "Henri de Gand et l'argument ontologique." *Archives d'histoire doctrinale et littéraire du moyen âge* 10 (1935-1936), 265-323.

_____. "Les disputes d'Henri de Gand et de Giles de Rome sur la distinction de l'essence et de l'existence." *Archives d'histoire doctrinale et littéraire du moyen âge* 15-17 (1940-1942), 323-358.

_____. "A propos de la théorie de la connaissance d'Henri de Gand." *Revue philosophique de Louvain* 47 (1949), 493-496.

_____. "Henry of Ghent." In *The New Catholic Encyclopedia*. New York: McGraw-Hill, 1967. Vol. 6, 1135-1137.

Pegis, Anton C. "Four Medieval Ways to God." *The Monist* 54 (1970), 317-358.

_____. "Henry of Ghent and the New Way to God (III)." *Mediaeval Studies* 33 (1971), 158-179.

_____. "A New Way to God: Henry of Ghent (II)." *Mediaeval Studies* 31 (1969), 93-116.

_____. "Toward a New Way to God: Henry of Ghent." *Mediaeval Studies* 30 (1968), 226-247.

Porro, Pasquale. *Enrico di Gand. La via delle propositioni universali*. Bari: Levante, 1991.

Rovira Belloso, José M. *La visión de Dios según Enrique de Gante*. Barcelona: Casulleras, 1960.

San Cristóbal-Sebastian, A. *Controversias acerca de la voluntad desde 1270 a 1300*. Madrid, 1958.

Schleyer, K. *Anfänge des Gallicanismus im 13ten Jahrhundert. Der Widerstand des französischen Klerus gegen die Privilegierung der Bettelorden. Historische Studien* XIV. Berlin, 1937.

Schöllgen, Werner. *Das Problem der Willensfreiheit bei Heinrich von Gent und Herveus Natalis. Ein Beitrag zur Geschichte des Kampfes zwischen Augustinismus und Aristotelismus in der Hochscholastik*. Düsseldorf: Schwann, 1927; repr. Hildesheim: Gerstenberg, 1975.

Smalley, Beryl. "A Commentary on the Hexaemeron, by Henry of Ghent." *Recherches de théologie ancienne et médiévale* 20 (1955), 60-101.

_____. "Gerard of Bologna and Henry of Ghent." *Recherches de théologie ancienne et médiévale* 22 (1955), 125-129.

Stadter, Ernst. *Psychologie und Metaphysik der menschlichen Freiheit. Die ideengeschichtliche Entwicklung zwischen Bonaventura und Duns Scotus*. München, 1971.

_____. "Die Seele als 'minor mundus' und als 'regnum.' Ein Beitrag zur Psychologie der mittleren Franziskanerschule." In *Miscellanea Mediaevalia*. Vol. 5: *Universalismus und Partikularismus im Mittelalter*. Berlin: 1968. Pp. 66-68.

Sweeney, Leo. "Divine Infinity: 1150-1250." *The Modern Schoolman* 35 (1957), 38-51.

Twetten, David B. "Why Motion Requires a Cause: The Foundation for a Prime Mover in Aristotle and Aquinas." In *Philosophy and the God of Abraham: Essays in Memory of James A. Weisheipl, OP*. Ed. R. James Long. Toronto: Pontifical Institute of Mediaeval Studies, 1991. Pp. 235-254.

Van Steenberghen, Fernand. *La philosophie au XIIIe siècle*. Louvain: Publications universitaires, 1966.

Wilson, Gordon A. "Thomas Aquinas and Henry of Ghent on the Succession of Substantial Forms and the Origin of Human Life." *Proceedings of the Catholic Philosophical Association* (1990), 117-131.

Wippel, John F. "The Condemnations of 1270 and 1277 at Paris." *The Journal of Medieval and Renaissance Studies* 7 (1977), 169-201.

_____. "Divine Knowledge, Divine Power and Human Freedom in Thomas Aquinas and Henry of Ghent." In *Divine Omniscience and Divine Omnipotence*. Ed. T. Rudavsky. Dordrecht: D. Reidel, 1985. Pp. 213-241.

_____. "Godfrey of Fontaines and the Act-Potency Axiom." *Journal of the History of Philosophy* 11 (1973), 299-317.

_____. "Godfrey of Fontaines and Henry of Ghent's Theory of Intentional Distinction between Essence and Existence." In *Sapientiae Procerum Amore: Mélanges médiévistes offerts à Dom Jean-Pierre Müller, O.S.B.*. Ed. T. Kohler. Rome, 1974. Pp. 289-321.

_____. "The Reality of Nonexisting Possibles according to Thomas Aquinas, Henry of Ghent, and Godfrey of Fontaines." *Review of Metaphysics* 34 (1981), 729-758.

Index of Proper Names

Index of Terms

MARQUETTE UNIVERSITY PRESS

Mediaeval Philosophical Texts in Translation

Published by: Marquette University Press, Marquette University, Milwaukee, WI 53233. **Manuscript submissions should be sent to:** Chair, MPTT Editorial Board, Dept. of Philosophy, Coughlin Hall, Marquette University, Milwaukee, WI 53233.